BOTANICAL CURSES AND POISONS

THE SHADOW-LIVES OF PLANTS

BOTANICAL CURSES AND POISONS

THE SHADOW-LIVES OF PLANTS

FEZ INKWRIGHT

STERLING ETHOS
New York

To Kit,
For sharing my passion for the beautiful and downright creepy, and without whose support and patience this book (and my life) would be half of what it is now.

An Imprint of Sterling Publishing Co., Inc.

Written and illustrated by Fez Inkwright
First published in 2021 by Liminal 11

ISBN 978-1-4549-5671-6

For information about custom editions, special sales, and premium purchases, please contact specialsales@unionsquareandco.com.

Printed in China

10 9 8 7 6 5 4 3 2 1

unionsquareandco.com

Book design by Kay Medaglia
Typeset by Francesca Romano
Cover design by Fez Inkwright

Contents

Thou art slave to fate, chance, kings, and desperate men,
And dost with poison, war, and sickness dwell;
And poppy or charms can make us sleep as well
And better than thy stroke; why swell'st thou then?
One short sleep past, we wake eternally,
And death shall be no more; Death, thou shalt die.

John Donne, *Death Not Be Proud*

INTRODUCTION

If you drink much from a bottle marked 'poison',
it is almost certain to disagree with you, sooner or later.

Lewis Carroll, *Alice in Wonderland*

There is no question that the evolution of the human species is closely intertwined with the plants with which we share our planet. As a foodstuff, they have strengthened us and carried us through times of poverty and famine. As a material, they have clothed us and built our shelters. As medicine, they have seen off disease and discomfort, and after that have grown over our graves. In our early religions, legends, and mythology, they've played a core role in shaping how we interact with our gods and environment.

But not all plants have made themselves friends to us. All children are warned, growing up, about the ones that sting, the ones that stick, and the ones we shouldn't put in our mouths. It is these hidden, unexpected dangers, and the idea that something so benign as plants can bring us harm, that has captured our imaginations since the beginning of time. Many of Shakespeare's plays embrace the dark drama of botanical poisonings, from *Hamlet*, to *Romeo and Juliet*, to *Anthony and Cleopatra*. The Greeks, too, filled their evenings with tales of Medea and Jason, in which the sorceress Medea helps Jason to find and win the Golden Fleece with the assistance of her knowledge of herbs and magic; so too with the tale of Heracles and Nessus, in which Heracles finally meets his death through trickery and a poisoned robe. At Alnwick Castle, UK, the Duchess of Northumberland was so inspired by a trip to the Medici poison garden in Italy that she created a garden of her own with plants that could kill instead of heal, with just one requirement for all of its residents: each one had to tell a good story.

We like to imagine that we know all of the usual suspects when it comes to the dangers in our gardens. Deadly nightshade. Wolfsbane. Nettles. Most of their common names do well enough in warning us away, such as baneberry, dogbane, henbane, and death cap mushroom. Some have, through their poisonous nature, become associated with the Devil of Christian legend; it's not uncommon to encounter names such as Devil's bit, Black Man's eye (the Black Man being a folk name for the Devil), or the Devil's claw, Devil's thread, candle, fingers… whatever the unfortunate plant may appear to resemble.

But not all plants are so blatant in their malignant natures, and these hidden dangers can make them just as lethal as their better-labelled cousins. Many gardeners have fallen foul of the cyanide fumes given off by laurel clippings being transported in a vehicle, and even the innocent-seeming potato and tomato, both common crop foods, can cause serious illness. Many of the plants that we encounter in our daily lives are, to some extent, toxic or otherwise harmful in some regard. Amongst the approximated 20,000 species of seed plants native to or naturalised in the United States, approximately 700 of these are known to have toxic characteristics, and elsewhere in the world, the number is higher still.

Perhaps it is our fault that we persist in coming into contact with plants that can kill. These toxic components have evolved out of a need to protect against consumption, after all, and yet we continue to consume; developing newer and more elaborate ways to avoid the unpleasant effects which are designed to stop us. Many popular starchy root plants, for example, contain cyanogenic glycosides which can be deadly if the roots aren't soaked, then drained, then soaked again multiple times over several days before grinding into flour. Some we don't even bother to process before eating: many people freely eat chilli peppers, even enjoying the burning sensation, unaware that they are a close relative of the deadly nightshade. Venomous animals and insects have their bright colours to warn predators of their dangers, but many plants rely on a bitter taste or a sharp sting instead. And for those too slow to take the hint, poison is a more effective—and usually more permanent—solution.

But poisonous does not always mean deadly. The official definition of a poisonous plant is 'that which contains substances capable of producing varying degrees of discomfort and adverse physical or chemical effects, or even death, to humans and animals when they are eaten or otherwise contacted'.* Not all of the plants included in this book are, by nature, poisoners, but are those that have evolved some other nefarious means of ensuring their survival. Among others you will find in these pages the gas plant (*Dictamnus albus*), which creates about itself a highly flammable atmosphere, starting wildfires and therefore burning out its

*Elizabeth McClintock and Thomas Fuller; *Poisonous Plants of California*

competition; and the strangler fig, which starts its life in the branches of other trees and then smothers its host so completely that the older tree rots away, leaving only an empty shell.

Here, too, are the unfortunate, the cursed, and the grieving of the plant world; those who are, to all appearances, harmless, but have throughout human history become associated with trolls, ghosts, murders, malicious spirits, or even the Devil himself. After all, who doesn't love a good ghost story or unresolved murder? The association between plants and the supernatural is long and tangled, and our reliance on the provisions of the earth—and our fear of its dangers—is still a strong, instinctive calling. The danger of 'fairy' mushroom rings has been taught to children for centuries, and there are even characters in popular culture that link back to these warnings in ways we might not expect: Shakespeare's famous Puck, for example, was based on the 'real' king of the fairies, whose name derived from the old English *pogge*—a toadstool.

Forests, old trees, and marshlands can be fearful places, and many fictitious creatures have sprung up across the world to explain the unease the wilder parts of our earth inspire in us. The Indian *Bhuta*, a restless ghost, dwells in trees waiting to possess the unwary. The *Siltim* is a similar creature which haunts the forests of Persia, modern-day Iran. An old Russian proverb claims that 'from all old trees proceeds either an owl or a devil', and in Germany, a country where agriculture has been a staple livelihood for centuries, the calendar and ritual of sowing and harvest has birthed generations of harvest goblins and

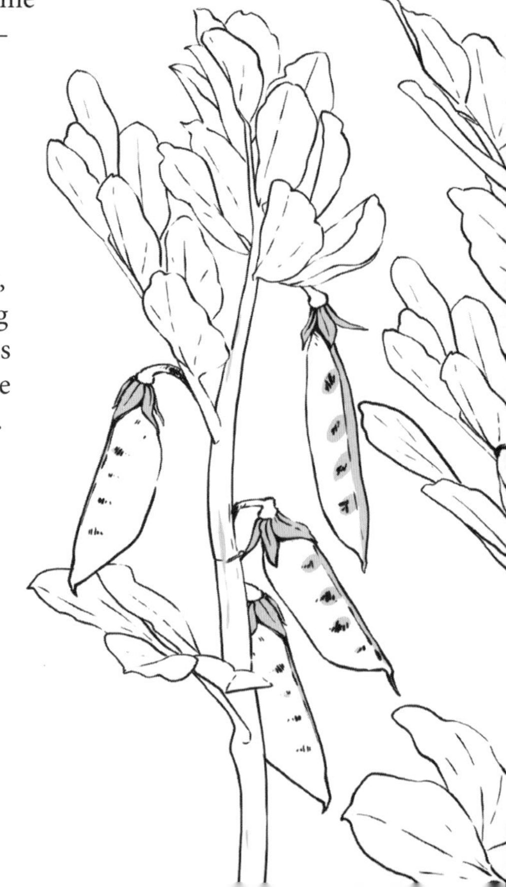

spirits to aid or hinder in the lives of farmers. Amongst dozens of others are *Aprilockse*, a spring field demon, *Gernstenwolf*, the barley-wolf, the *Graswolf* in the fields, *Ernetbock*, a demon who steals corn from the harvests, and *Heukatze* and *Heupudel*, who bring blight to hay stored in barns.

Though talk of demons and goblins evokes images of a wilder, more primitive world, many of these tales and their wisdom persisted well into the 1900s, and still exist in some forms in our lives today. After all, it's still common knowledge in many countries that tossing salt over your shoulder will protect you from the Devil, and saluting a lone magpie will ward off bad luck. Whimsical it may seem, but our cultural make-up is that of farmers and country folk, and it turns out that centuries of superstition are hard to be entirely rid of. In the late 1800s and early 1900s there was a resurgence in academic interest in folklore, particularly in the British Isles, where books and journals recording local beliefs were researched and published in huge numbers. Though these provide a remarkable insight into rural life at the time, much of the content was based on tales gathered directly from the source, this being locals and not academics, and as a result many of the stories gathered display a rather fantastical concept of magic. Due to these, it's worth taking some of the stories in this book—and mentions of those mystical people known as Druids and pagans—with a grain of salt.

Though Druidism still exists today as a spiritual movement, mention of Druids in old writings, particularly those that deal with local lore and discussions of the occult, are often exaggerated and inaccurate. Historically, Druids were a part of the Insular Celtic (within the British Isles) religion, or were practitioners within the historical region of Gaul,

which covered large parts of Europe. Druids were teachers, scientists, philosophers, and above all priests, but the tales of mysticism and power that surrounded them have led to hundreds of stories, most of them inflated for dramatic effect. So though the idea of golden scythes and sacrifices of white bulls may make for great stories, we have to keep in mind that they may be concepts broadly applied to the romantic concept of 'Druidism' as a magical art, and not entirely historically accurate.

The same applies to the catch-all term 'paganism', which was spread by the early Christian fear of magic and Devil worship, particularly during the period of the witch trials of the 1500s. Originally deriving from an insult used by late Roman Christians against rural people who continued to worship traditional gods instead of embracing the Church, 'pagan' was never actually a professed religious identity, but nonetheless came to define a loose system of religious beliefs and ways of life, the practice of which varied widely across ancient and medieval Europe.

Inaccuracies or exaggerations though some may be, at their heart, most folk tales exist as warnings to the unwary. Don't go into the forest at night, don't talk to strangers. Don't disrespect the land you live on; treat it kindly and you will be rewarded. Our dependence on the provisions of the earth is undeniable, and many early belief systems sanctified the plants that formed such a core part of their livelihoods. It is only fitting, then, that such a wealth of stories regarding this relationship between man and nature can be found all over the world; and while a good deal have survived as much through oral means as written, they tell key truths about the people of the time, what they valued, and how they lived. This concept is summed up perfectly by folklorist Christina Hole:

The folklore of a nation must always be a matter of great importance to those who desire to understand the nature and history of its people.

This book compiles but a handful of these tales.

DISCLAIMER:

This book is written solely for informational and entertainment purposes. It is not intended as a source of medical advice, nor are the illustrations to be used as an accurate reference for identification. Please seek the advice of a professional medical practitioner before using plant-based medicine.

... we rest not contented with natural poisons, but betake ourselves to many mixtures and compositions artificial, made even with our own hands. But what say you to this? Are not men themselves mere poisons by nature? For these slanderers and backbiters in the world, what do they else but launch poison out of their black tongues, like hideous serpents?

Pliny, *Natural Histories*

A HISTORY OF POISONING

Long before guns, bombs, and the popularity of toxic chemical elements such as arsenic and mercury, the easiest way to be permanently rid of a problem was to take advantage of what nature provided. From Cleopatra's poisonous asps to the demise of Alexander the Great and the Roman Emperor Augustus, the natural world has been as apt a provider of deadly weapons as any we've been able to design ourselves. And much as history is full of tales of war and assassination, many of the most fascinating and memorable murders have come by way of poisonings. With the right poison and the right method of application, being able to dispose of a troublesome rival (or even spouse or parent!) was once an invaluable skill.

As a species, we've historically made ourselves apt at the art of murder, particularly in pursuit of power and self-advancement. There are cases of unsolved poisonings throughout our early history, with one even to be found in the Old Testament of the Christian Bible, regarding the death (supposedly by stroke) of the High Priest Alkimos in 159BC. The report on his death in the *Septuagint* (the earliest existing Koine Greek translation of the scriptures) presents the typical symptoms of apoplexy: collapse and loss of speech, followed rapidly by death. But they also record the High Priest as having suffered severe pains prior to his demise; an unusual occurrence in stroke patients, but remarkably similar to symptoms of the aconitine poison found in monkshood (*Aconitum napellus*). Not only was aconite a common and easily-cultivated poison of the time, but just before his death Alkimos had lost popularity by sanctioning construction work on the Temple of Jerusalem; a project considered sacrilege by many around him. Though it's still commonly considered that his death was divine punishment, academics are beginning to consider that there may have been a more mortal hand involved in his untimely demise.

You must remember that poisoning is always concealed and deliberate. It is a crime that is not done in a moment of passion, or on an impulse. It is a crime that must be planned.

Justice William Windeyer, Central Criminal Court Sydney, summing up the Dean case, April 6, 1895

One of the most famous early poisonings is that of Socrates in 399BC, by way of execution by hemlock. Accused of corrupting the youth of Athens and refusing to recognise the gods of the state, the famous philosopher and his anti-democratic ways were declared to have influenced two of his students to overthrow the Athenian government, leading to two brief but riotous periods which resulted in thousands of citizens being banished from the city or swiftly executed in order to restore the city's democratic laws. After a closely-contested trial that lasted twelve hours, it was decided that Socrates must die for his part in inspiring the unrest, and at his own hand too. Given a cup of hemlock extract, he would be expected to drink it and succumb to its effects. The best account of this scene comes from Plato, one of his students, who discussed Socrates' death and the subject of the immortality of the soul in great detail in his dialogue *Phaedo:*

When he saw the man [the executioner], *Socrates said: 'Well, my friend, you're an expert in these things: what must one do?'*

'Simply drink it,' he said, 'and walk about till a heaviness comes over your legs; then lie down, and it will act of itself.' And with this he held out the cup to Socrates.

...

He [Socrates] *walked about, and when he said that his legs felt heavy he lay down on his back—as the man told him—and then the man, this one who'd given him the poison, felt him, and after an interval examined his feet and legs; he then pinched his foot hard and asked if he could feel it, and Socrates said not. After that he felt his shins once more; and moving upwards in this way, he showed us that he was becoming cold and numb. He went on feeling him, and said that when the coldness reached his heart, he would be gone.*

By this time the coldness was somewhere in the region of his abdomen, when he uncovered his face—it had been covered over—and spoke; and this was in fact his last utterance: 'Crito,' he said, 'we owe a cock to Asclepius: please pay the debt, and don't neglect it.'

'It shall be done,' said Crito; 'have you anything else to say?'

To this question he made no answer, but after a short interval he stirred, and when the man uncovered him his eyes were fixed; when he saw this, Crito closed his mouth and his eyes.

And that, Echecrates, was the end of our companion.

Plato, *Phaedo*, translation by David Gallop

Poisoning was a prevalent means of assassination during Socrates' time, being a convenient way to remove a political opponent, an unwanted spouse or step-child, or even ensure an earlier inheritance by removing an elderly parent. And with poisonous plants such as aconite, colchicum, henbane, mandrake, hellebore, poppy, and yew easily available in most gardens or growing wild, it wasn't just convenient but also cheap and easily accessible.

The first recorded instance of mass poisoning in Rome was in 331BC. Though the large number of deaths were initially passed off as the result of an epidemic, a slave girl passed information to elected officials that the deaths were the result of poisons prepared and administered by Roman women, and upon investigation, about twenty matrons—most wealthy land owners—were discovered in the process of creating poisonous mixtures. Though they claimed these concoctions were harmless, they were forced to drink them to prove themselves innocent, and promptly died. In later investigations, a further one hundred and seventy more were found guilty of the same offence and executed.*

Almost one hundred and fifty years later in 184BC came another case of mass poisoning, in connection with the worship of Dionysus, the Greek god of wine, ritual madness, and religious ecstasy. Female followers of Dionysus were known as *maenads,* and would intoxicate themselves by chewing ivy leaves, which can cause madness and anger. They would then begin a drunken rampage across the countryside, attacking animals and humans alike. An offshoot of this cult caused such trouble that the praetor Quintus Naevius spent a vast amount of public funds on a four-month investigation into the affair, resulting in the trial and execution of two thousand people under the main charge of poison.† Further executions were carried out four years later in 180BC, as officials tried to curb this substantial threat to Roman society.

In 82BC, poisoning had become such a regular occurrence in Rome that the general and statesman Sulla declared it a capital offence. It became illegal to make, buy, sell, possess, or to give poison for the purpose of killing (though it remained legal to obtain it for pest control and medical practices), under punishment of deportation and the confiscation of property. Aconite was such a popular garden plant—both for its beauty as a flower and its more practical use—that it was given a particular mention in these laws. However, Sulla's attempts appear to have had little effect on the popularity of the act, as 81 years later in 1BC the popular satirist Juvenal, remarking upon the moral decay of the elite, claimed that poisoning for personal benefit had become something of a status symbol.

*David B. Kaufman; *Poisons and Poisoning Among the Romans*
†Livy; *Ab Urbe Condita (The History of Rome)*

Shall I speak of spells and love-potions too, poisons brewed, and stepsons murdered? ... Greed is usually the root of crime: no fault of the human mind causes more poison to be mixed, or a more frequent rampaging about with a blade than the uncontrolled desire for extravagant wealth. For the man who wants to be rich, wants to be rich now; but what reverence for the law, what fear or shame can you expect from a greedy man in a hurry?

Juvenal, *The Satires*

Murder by poison certainly features heavily in the history of Roman Emperors and the tumultuous rise and fall of their Empire. One famous political player was Locusta, a notorious poisoner responsible for the assassinations of multiple high-profile targets, including Emperor Claudius, who died by way of poisoned mushrooms in 54AD.

Locusta was a Gaul who, alongside Canidia and Martina, became one of an infamous trio of woman poisoners, or *venefica*, meaning 'practicing both poisoning and sorcery'. The word *veneficus* or *venefica* was applied to a poisoner, or maker of poisons. Not much is known of Locusta's early life, but she arrived in Rome with a deadly knowledge of herblore, and kept in her arsenal hemlock, foxglove, nightshade, and opium. Testing her extracts on animals, she honed them with a lethal and scientific efficiency and, though she ended up imprisoned at least twice in her career, each time found freedom again through the influence of her wealthy benefactors, who had need of her particular skills. Tacitus, in his Annals, describes her thus:

'This was the famous Locusta; a woman lately condemned as a dealer in clandestine practices, but reserved among the instruments of state to serve the purposes of dark ambition.'

For some time, Locusta was in the employ of Empress Agrippina the Younger, Claudius's niece and then-wife. As a matter of fact, Locusta was already imprisoned on earlier charges of poisoning when Agrippina approached her to assassinate Claudius; her purpose being to make way for Nero, Agrippina's own son from a previous marriage. Nero himself later employed Locusta to do away with his step-brother, Claudius's son, Britannicus. In exchange for doing this, Locusta was rewarded with a full pardon and a country estate, where pupils were sent to learn the craft from her. Nero went on to keep a personal poisoner, but preferred cyanide to the slower-working tropane alkaloids that Locusta employed. He maintained one last use for her, however: when it came time for him to flee Rome in 68AD, he acquired from her a poison for his own use, should he ever need it, though eventually he died by other means.

It wasn't just the Romans who made good use of poison. Mithridates VI, the King of Pontus between 114BC and 63BC, feared death by poison so acutely that throughout his life he took minute daily doses of them to build up an immunity. When finally captured by the Romans in the Mithridatic Wars in 63BC, rather than be taken alive he tried to poison himself; but, rather unsurprisingly by this point, survived.

Mithridates spent many years over the course of his reign applying poisons to those on death sentences to test his antidotes. As with most medical practices at the time, these experiments were heavily influenced by religion: Mithridates kept a contingent of Scythian doctor-shamans with him at all times, who supervised many of his studies, and were a distinct influence on the King's work. They came from the Agari tribe north of the Sea of Azov, in modern-day Ukraine, and were experts in poisons and anti-poison methods alike. At one point they supposedly saved his life on the battlefield by applying snake venom to a wound in his thigh to staunch the bleeding, and, though feared as mystical northmen by the rest of Mithridates' court, were valuable for their knowledge of many highly poisonous plants and their applications.

The land of Pontus was by no means lacking in resources to help Mithridates in his studies, either. Bees that fed on the nectar of oleander and rhododendrons created wild honey thick with deadly neurotoxins, and beavers that fed on willow were prized for their flesh high in salicylic acid. Pontus's eastern ally, Armenia, boasted lakes full of poisonous fish and snakes. Ducks ate a diet that included hellebore and nightshade with no detriment to their own health, and were the source of an ingredient which Pliny in particular noted as being used in Mithridates' work: 'the blood of a duck found in a certain district of Pontus, which was supposed to live on poisonous food, and the blood of this duck was afterwards used in the preparation of the Mithridatum, because it fed on poisonous plants and suffered no harm.'

Mithridates eventually created a formula known as *Antidotum Mithridaticum*, a general antidote against many common poisons which was so successful that even after his death the Romans took and translated it, and continued to use it after his defeat by Pompey the Great. The only remaining recipe that we have for *Antidotum Mithridaticum* is the one recorded by Pliny. As well as comprising dozens of the best-known herbal remedies from the time, it lists 54 poisons in small amounts. Though accounts maintain that the antidote was genuine—even becoming so well known that for some time, *Mithridate* became the common word for any antidote—none of the ingredients listed by Pliny are particularly known for being effective against poisons (aside from a few mild purgatives, such as rhubarb root), and it remains a topic of debate amongst historians today as to whether its restorative properties were a genuine claim, or a fictional one. It may even be that tales of the panacea's effectiveness were spread by Mithridates himself to hide his real secret: the increased resistance to poisons that he gained through daily exposure.

Regardless of its veracity, the tale has become a famous one, even so much as to feature in poems such as the following from the renowned Alfred Housman:

There was a king reigned in the East:
There, when kings will sit to feast,
they get their fill before they think
Of poisoned meat and poisoned drink.
He gathered all the springs to birth
from the many-venomed earth;
First a little, thence to more,
he sampled all her killing store
And easy, smiling, seasoned sound
sate the king when healths went round.
They put arsenic in his meat
and stared aghast to watch him eat;
They poured strychnine in his cup
and shook to see him drink it up:
They shook, they stared as white's their shirt:
them it was their poison hurt.
I tell the tale that I heard told;
Mithridates, he died old.

A. E. Housman, *A Shropshire Lad LXII*

By the 16th Century, the use of toxic chemicals such as arsenic had risen in popularity, particularly in Europe. The symptoms of arsenic ingestion are similar to those of cholera, which was a common illness at the time, and as such it provided the perfect way to remove troublesome persons. No doubt the popularity of poisoning continued for all the same reasons that the Romans employed, as by the 19th Century arsenic would become colloquially known as 'inheritance powder'.

Most infamous for such schemes were houses Borgia and Medici, two prominent Italian families which produced between them five popes and two regent queens of France, and both of which were the subject of suspicion in many crimes. The Borgias gathered enormous wealth by abusing laws that saw the properties of their victims reverted to the church (and therefore to them); and the Medicis—most notably Ladies Catherine and Marie de Medici—were rumoured to have a room with 237 tiny cabinets of poisons hidden within its walls. Catherine in particular, wife to one French king and mother to three more, was wont to meddle in affairs of state and was implicated in more than a few mysterious, yet convenient, deaths.

By 1531, Henry VIII declared 'poisoning as a wilful act of murder' to be an act of treason, with all of those accused punishable by being boiled alive. This may partially have been a reaction to the widespread political assassinations occurring across Europe, as King Henry himself had a great fear of befalling such a fate, and poisoning is also thought to have been responsible for the death of his former wife, Catherine of Aragon. However, it is likely that it was a law created for his own convenience: at the time of passing, a cook by the name of Richard Roose sat in jail, whom Henry is suspected of hiring to assassinate Bishop John Fisher. Fisher was Henry's one-time tutor, who now stood against the King in matters of state, and Roose was hired to poison his meal. On the night of the event, however, Fisher was feeling too unwell to eat and instead two servants partook of the unwanted broth. Arrested for their deaths, Roose could neither provide a reason for why the meal had been poisoned and nor could Henry afford to let him go free. *An Acte for Poysoning* was pushed through parliament with haste, and Roose was tried as though he had killed a royal rather than two servants who had, by chance, happened to partake of the meal. He stood trial, was found guilty, and subsequently executed by boiling, all within six weeks of the crime.

Only three people were ever subjected to this unusual punishment: Richard Roose; an unknown maidservant from King's Lynn; and Margaret Davie in 1542, who had poisoned the occupants of all three households that she worked for. The act was later repealed in 1547 by Henry's son, Edward VI, six years before he died… supposedly of poisoning.

O, brothers mine, take care! Take care!
The great white witch rides out to-night.
O, younger brothers mine, beware!
Look not upon her beauty bright;
For in her glance there is a snare,
And in her smile there is a blight.

James Weldon Johnson, *The White Witch*

OF WISE WOMEN AND WITCHES

After Europe's poisoning spree of the 1500s, the story of humanity and its love of assassination moved more towards chemical and mechanical warfare. But for one group of people in particular, accusations of 'poisoner' persisted for many more years to come, and with horrifically brutal results. To understand the full implications of this, however, we must first begin even a few centuries earlier.

The 1300s ushered in three and a half centuries of fanatical hatred towards herbalists, mostly female, who were seen as 'touched' or otherwise gifted with healing. This massive and prolonged crusade was sanctified by both the Christian Church and the government (which at this time was so entangled with the Church as to be practically the same entity), and created a hysteria that saw an estimated 63,850 witches (as officially recorded, not including those uncounted who fell afoul of vigilante justice) burned, drowned, hanged, and crushed.

Nothing aroused greater wonder among the ancients than botany. There still exists a conviction that these phenomena [the phases of the moon] are due to the compelling power of charms and magic herbs, and that the science of them is the one outstanding province of women.

Pliny, *Natural Histories VII*

But even in the 1300s, witches weren't a new concept to the world. Pliny wrote often of local 'wise women' who were well known, and indeed sought out, for their ability to either cure or curse. Witches were devoted to Hecate, the Grecian goddess of Hell, who presided over magic and enchantments. Hecate's daughters, Circe and Medea, were famous for their mystic pharmacopoeia of herbs—particularly poisonous ones.

Botanical poisons, it seems, have gone hand in hand with women and witches ever since early Roman records and likely long before this still. A weapon for the weak against the strong, an invisible and untraceable source of fear, poison has been the supposed weapon of women against menfolk since time immemorial. Reginald Scot, in *The Discoverie of Witchcraft* (1584), noted that 'Women were the first inventers and practisers of the art of poisoning', which they are 'more naturallie addicted and given thereunto than men'. Even so late as 1829, Robert Christison wrote in his *A Treatise on Poisons*: 'The art of poisoning, in all ages of the world, has been chiefly indebted to the female sex for its scientific cultivation.'

Women, particularly Pliny's 'wise women', were both to be respected and feared. They had the knowledge to kill, but also the knowledge to heal; they knew local plants better than anyone else, and across Europe took on the role of midwife, nurse, and even seer, reading the skies and predicting the next day's weather. All of this knowledge, passed down through generations, almost exclusively by spoken word, rarely written down; and none of it sanctioned by an official education nor—most importantly to their eventual persecution—the Christian Church.

Thou shalt not suffer a witch to live.

Exodus 22:18

The Church stood in judgement of all outsiders, and viewed these women healers—who were not sanctified by God as their own doctors were—as a threat to their authority. The intense research into early medicine that Greek and Roman physicians had undertaken had been

all but forgotten by the educated world, and medical services were provided in monastic hospitals that sprung up across the continent. However, the care provided in these was rudimentary at best, and more often than not was simply palliative.

It wasn't until the 1200s that new translations of older texts began to re-emerge, providing medical schools with the knowledge that they had been lacking. But by this late point, Church-sanctioned colleges and doctors were far behind the skill of those wise women who had continued to practise the knowledge of their forebears for all these long years.

The persisting capability of hedge witches—and the faith that the general populace put in their healing abilities—was a threat to the Church's authority. A pointed and vicious propaganda campaign was mounted against them, turning public opinion and proclaiming the dangers of these women in every church across Europe. Best known, perhaps, is the corruption of the famous line from Exodus: *Thou shalt not suffer a witch to live.* This translation, a mantra that is held responsible for much of the witch madness that swept the continent, lives on in most Bibles even today. But the original Hebrew word in this passage is *mekhashepha*, a word translated by the *Septuagint* as *pharmakeia*: poisoner. This simple and convenient 'new' translation gave the early witch hunters every excuse they needed to spread their hatred. The persecution of witches was now sanctified by God, and they had every reason to begin their hunt.

Until the 1200s, the Church had preached that illnesses were inflicted by God as a punishment for sin, but now the Inquisition changed their doctrines to declare that illness, particularly the kind which lay beyond the ability of their own physicians to cure, must be the creation of sorcery. Witches were declared to be agents of the Devil, usurpers of the powers of God. They provided miracles that the

Church's doctors could not yet match, directly contradicting the Bible: *Blessed be the Lord God, who* alone *doeth wondrous things.* God simply could not be the only one capable of wondrous things if these witches, too, had access to some great power.

One of the men responsible for the greatest condemnation of witches was Henri Boguet, the Grand Judge of St Claude, France. His publication, *Discours Exécrable des Sorciers,* was so popular that it was reprinted twelve times over twenty years, and by 1590 he alone had ordered the executions of six hundred women whom he saw as 'the deadliest enemies of Heaven'. He claimed that their 'cures' only lead to further illness so that they might maintain their power over men, and it was largely due to his influence that the hysteria spread so far across Europe. As he observed in his writings: 'Germany is almost entirely occupied with building fires for them. Switzerland has been compelled to wipe out many of her villages on their account. Travellers in Lorraine may see thousands and thousands of the stakes to which witches are bound...'

The accusations made against wise women and rural healers found their roots in superstition, and were largely without merit. It was believed that, amongst their other numerous talents, they could turn men against each other, infect cattle, cause storms, and make women barren.* The only thing that they could not do was directly kill a man: but with poisons gifted to them by the Devil (who, according to Boguet, had a knowledge of every plant on the earth), even that great crime was within their repertoire.† Even so late as the 17th Century, midwives were declared to be 'especial favourites of the Devil', as their mystical knowledge of birth and the female body was not believed to have been mortally possible.‡

More recent theories have posited that the witch trials were a form of propaganda during the 1517 Reformation when the Church was separated into two factions, Catholic and Protestant. After a sustained period of crop failure across the continent and an ecological

* Robert Burton; *The Anatomy of Melancholy*

† George Gifford; *A Dialogue Concerning Witches and Witchcrafts*

‡ Christian Stridtbeckh; *Concerning Witches, and those Evil Women who Traffic with the Prince of Darkness,* 1690

disaster known as the Little Ice Age, the idea of finding a scapegoat to blame for these hardships seemed a sound idea: but with the added bonus of being able to remind people of the protection that the Church and its new branches could provide against this threat.

Though the majority of witch hysteria occurred across Europe, news and superstition travels fast, and elements of fear still spread further. Some of the last witch trials of that time occurred in 1692, and saw nineteen executed in Salem, Massachusetts, after a period of hallucinations and illness that struck the town. The accused were mostly elderly women, almost all of them widows, or of frail physical and mental health. Though it is generally now thought that the symptoms experienced in Salem were the result of ergot-infected bread, these women were tried and found guilty of poisoning and of corrupting the town, and subsequently hanged. As with the events in Rome in 331AD which saw 2,000 supposed witches executed, this is a key example of another mass-poisoning leading to witch hysteria.

WITCHING HERBS

Once you have tasted flight, you will forever walk the earth with your eyes turned skyward, for there you have been, and there you will always long to return.

Leonardo da Vinci

One of the best-known accusations laid against witches was the ability to fly, in order to attend the sabbats regularly held by the Devil. These sabbats were said to involve all manner of wild revels, including 'the attendants riding flying goats, trampling the cross, and being re-baptised in the name of the Devil while giving their clothes to him, kissing his behind, and dancing back to back forming a round'.* Many modern stereotypes about historical witches still mention these naked, dancing orgies, but what 'official' records we have of them were mostly taken during the period of the witch trials, quoted from witches confessing under torture and written by priests who had never taken part in these gatherings. Ultimately, they were little more than an effective piece of propaganda to promote the perceived wickedness of the accused.

But the belief that witches were able to fly—by the aid of a goat or a broomstick or whichever cited method—still pervades. So where does it come from? Rather than actual, physical flight, witches captured during the trials were observed to use a 'flying ointment', which was a compound of various psychoactive plants that would have caused the sensation of flying. The ointment wasn't a new invention; this green salve is mentioned in Homer's *Iliad,* and in the novel *The Golden Ass* by Apuleius, a Roman tale from the 2nd Century in which a witch uses an ointment to transform herself into an owl. Earlier still, in 800BC, Queen Hera used a mixture called the 'oil of Ambrosia' to fly to Olympus, which may have been the very same mixture.

In these stories the ointment was certainly never connected with the Devil, and it seems to have been forgotten by history until 1324, when it re-emerged in the trial of the Irish witch Alice Kyteler, suspected of poisoning her fourth husband. Those who arrested Alice found in her home 'a wafer of Sacramental bread, having the Devil's name stamped thereon instead of Jesus Christ, and a pipe of ointment wherewith she ambled and galloped through thicke and thin…'†

The art of flying, or transvection, was never voluntarily mentioned by the accused during during the hearings that had been written down by clergymen, but was of great interest to the Church and its witch hunters. Most of the recipes that were recorded during the hearings had been written by clergymen (such as in Reginald Scot's *The Discoverie of Witchcraft)*, or by later scholars (a good example is

*Francesco Maria Guazzo; *Compendium Maleficarum,* 1608
† St John Seymour; *Irish Witchcraft and Demonology*

botanist William Coles' *The Art of Simpling* in 1656), and therefore most of formulae that we know of now are likely inaccurate in some form or another. However, there are similarities in the ingredients and techniques employed, and so a general idea of the salve's contents can be deduced.

The ointment was typically made from a base of fat or oil, darkened with soot, and may have contained poisons such as deadly nightshade, mandrake, datura, henbane, wolfsbane, and hemlock. All of these plants grow freely in Europe and could have been easily and freely gathered; furthermore, they are ingredients which are easily absorbed, even through unbroken skin, and can produce impaired movement, irregular heartbeat, dizziness, and excitement. Two of the aforementioned plants in particular, wolfsbane and deadly nightshade, are notable for their disruption of the heart, as discussed in Margaret Murray's *The Witch-Cult in Western Europe*: 'irregular action of the heart in a person falling asleep produces the well-known sensation of suddenly falling through space, and it seems quite possible that the combination of a delirifacient like belladonna with a drug producing irregular action of the heart like wolfsbane might produce the sensation of flying.' In the early 1900s, German folklorist Dr Will-Erich Peuckert experimented with this mixture of soot, plants, and fat, and experienced first-hand some of the sensations. He wrote: 'We experienced in dreams, first wild and yet restricted, flights and then chaotic revels like the wild tumult of an annual fairground, and finally progressed to an erotic licentiousness.'

The toxins contained particularly in the nightshade family can also produce the sensation of turning into an animal, as witches supposedly had the ability to do. Reports from those accidentally poisoned by plants from this family have described the afflicted imagining themselves to grow fur, ears, or feathers.* Giambattista della Porta, a 16th Century scholar, discussed these in his most famous work *Magiae Naturalis* (*Natural Magic*), which described in great depth the ointments, in particular one that caused extreme thirst, impaired vision, and a staggering gait; all common signs of lycanthropy. All of these are also, coincidentally, symptoms of atropine, which is present in henbane, datura, and deadly nightshade.

We know of several instances where witches were observed on these sabbatical flights. The first was recorded by della Porta, regarding his study of a witch who slept for some time, and was unable to be awakened. Court physician Andrés Fernandez de Laguna writes of another who anointed herself from head to toe before him, and slept for 36 hours continuously, despite attempts to wake her. When

* Claire Russell and William Moy Stratton Russell; *The Social Biology of the Werewolf Trials*

she finally came to, she declared she had been 'surrounded by all the pleasures and delights in the world', including the attentions of a 'lusty young man'. The final incident is the observation of a French doctor, who observed the flights of several women. One Bordeaux witch slept for five hours, and when she awoke told the doctor correctly of several incidents that had occurred whilst she slept. Seven more women performed a three-hour trance before him, and told him after of several verifiable events that had happened within ten miles of their location. For their confession, they were burned.*

* Jennifer Bennett; *Lilies of the Hearth: The Historical Relationship Between Women and Plants*

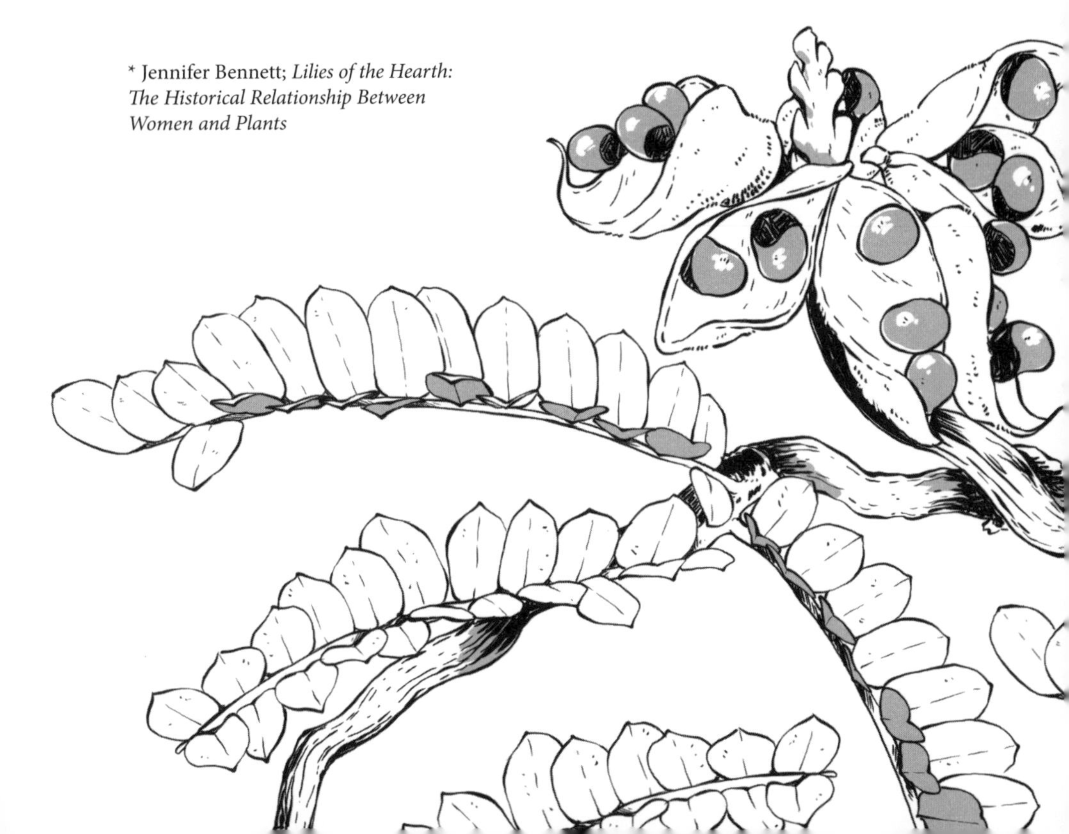

All things are poison, and nothing is without poison, the dosage alone makes it so a thing is not a poison.

Paracelsus

TO CURE AND TO KILL

While many of the plants mentioned thus far are noted by history for their lethal properties, it would do them a disservice not to mention the role they have played in the field of medicine. Though our knowledge of medicine, dosages, and safer alternatives has increased significantly in more recent centuries, these dual-natured plants, which possess properties both beneficial and malign, have undoubtedly earned their place in history.

Death and healing commonly walk hand in hand, and many of these plants were seen as representative of these dual realms, assigned to the early deities and spirits who walked the same line. And across the world, there are many of them. The Sumerian goddess Gula is called the Great Healer; yet she also curses wrongdoers with poisonous herbs. Omolu, the divine vodou plague-doctor (also known as Babalú-Ayé, or Sakpata in the Yoruba religion), is a god of both contagion and deliverance from disease. This duplexity is even mirrored in the Greek word *pharmakon,* from which we get *pharmacy,* which means both 'cure' and 'poison'.

Some plants were used—provided that the correct dose was administered—as very early anaesthetics, while others took a role as early purgatives (both upwards and downwards), to empty the stomach and bowels of unwanted contents. The more poisonous, the more bitter the taste; but this was easily overcome with sugar or honey, and many patients underwent gruelling treatments in the name of a cure—the outcomes of which were sometimes as much a cause of their ultimate demise as their illness may have been.

One of the unfortunate recipients of such a regime was George Washington, when he was struck down with a sore throat and fever in 1799. Had he been left alone he may well have recovered, but reports state how overzealous doctors bled him of four and a half pints of blood (the average body containing ten pints in total), and then gave him three doses of mercury chloride and a purging enema. A further several doses of tartar emetic for purging were prescribed, and a blistering compound was applied to his throat and feet, with the intention that the blisters would draw out the harmful elements causing the sickness. Already weakened from illness, Washington died less than 24 hours after the doctors arrived.

Purges were dangerous, especially when prescribed to an already-sick man, but the use of them was rampant in centuries past. The *Syon Abbey Herbal,* written in 1517, suggests to 'take Black Hellebore, Ramsons, Henbane, vinegar, White Bryony, and mix them together with old grease. When required, rub the ointment onto the patient's hands and feet.' These ingredients would be absorbed through the skin and would help induce a purge. Some plants were even said to be able to control the direction of the purge—an Appalachian belief was that if the leaves of boneset (*Eupatorium perfoliatum)* were stripped upwards, they would inspire an upward purge; but downwards would make them instead a cathartic. In reality, boneset purges both ways, no matter which way the leaves are removed from the stem.

Poisonous juices, used in modest amounts, were also valuable as an anaesthetic, albeit an unreliable one. One of the earliest-known medical sedatives was in use 2,000 years ago, and consisted of a combination of hemlock, mandrake, and henbane. The juices of these plants would be soaked into a sponge which would then be dried out and, when needed, soaked in hot water to create a steam for the patient to inhale.* This soporific sponge was called a dwale, from the Danish *dvale*, meaning a deadly trance, and when it entered the English dictionary also came to be an alternative term for an opiate. In Fala, Scotland, a medieval hospital and church known as Soutra Aisle was excavated in 1986. In the cellar, the seeds of hemlock, henbane, and opium poppy were found, likely intended to be used as an anaesthetic in the same manner.

DOCTRINE OF SIGNATURES

> Though sin and Satan have plunged mankind into an ocean of infirmities, yet the mercy of God, which is over all His workes, maketh grasse to growe upon the mountains and herbes for the use of men, and hath not only stamped upon them a distinct forme, but also given them particular signatures, whereby a man may read even in legible characters the use of them.
>
> William Coles, *The Art of Simpling*

The doctrine of signatures was a medical concept developed by Paracelsus in the early 1500s, which theorised that plants resembling particular parts of the body must be useful in treating ailments of the corresponding anatomical area. Christian theologians, who quickly caught onto the concept, claimed that God must have created these signs to show humans how the plants could be useful to them. Similarities would be read into the shape of the specimens, the patterns of their leaves, and even their colours. Red plants were decided to be good for the heart, yellow for the spleen, green

* Stephen Pollington; *Leechcraft: Early English Charms, Plant Lore, and Healing*

for the liver, and black for the lungs. The brain-like shape of the walnut made it ideal for headaches and dizzy spells; beans were good for the eyesight; sunflower seeds would help with tooth pain; and strawberries were assumed to be an effective cure for heart conditions. Many of these assumed powers live on in the common names we still use today, such as in the lungwort, the liverwort, and eyebright. Imagination, rather than careful experimentation, became the favoured way to identify plants with healing properties. In reality, if they worked, it surely must have only been via luck or placebo.

Unfortunately, those promoting the doctrine listed a vast number of plants inappropriate for purpose, such as in the case of the birthwort (*Aristolochia spp)*. Due to its similarity in shape to the womb and ovaries, it was decided that it must be an ideal cure for health conditions relating to childbirth, particularly in expelling the placenta after birth. Unfortunately, birthworts are highly toxic, and can cause severe vomiting, kidney failure, and death; so even though it was administered with the best of intentions, it was more likely causing the miscarriage of children and the death of expectant mothers.

Unsurprisingly, the doctrine fell out of popularity in the 1800s, although it still continues to be used in some areas today as a form of folk medicine in some areas.

KALLISTI

A-Z of Plants

APPLE: Malus domestica

I will find out where she has gone
And kiss her lips and take her hands;
And walk among long dappled grass,
And pluck till time and times are done
The silver apples of the moon,
The golden apples of the sun.

William Butler Yeats, *The Song of the Wandering Aengus*

The humble apple is a popular fruit across the world, but it has also been associated with the dead, the underworld, and demonic possession for thousands of years. Just think of Snow White, or Adam and Eve, and you'll realise that the idea of the poisoned apple is a prominent motif in stories across the ages. The idea of this fruit being deadly isn't an entirely fictional imagining, either: the pips contain, albeit only in a mild quantity, cyanide.

Famous for smelling of almonds (although only 50% of people can detect the scent), cyanide occurs naturally in almond, cherry, and peach kernels, in addition to apple pips. The pips don't contain enough cyanide to cause an accidental overdose, however there is an amusing anecdote about English cider that suggests otherwise. Cider is made across the UK, but there are two regions—Norfolk and the West Country—that are particularly famous for it. Norfolk apples are pulped for their juice, but in the West Country they are crushed between millstones, supposedly causing trace amounts of cyanide to be present in the drink. Since part of an agricultural worker's wages used to include a gallon of ale or cider per week, it was said that workers in the West Country eventually went blind and mad from drinking it; although this anecdote was originally recorded as told by a Norfolk cider maker, so it may simply be an attempt to slander the opposition!

The association between apples and the themes of death and tragedy can be found throughout Greek mythology. The apple tree is supposed to have originated with Melos, a young man who met and fast became besotted with Adonis, the son of King Kinyras. When Adonis met his untimely death on a hunting trip, Melos was so taken by grief that he hanged himself from the branches of a barren tree. Touched by his tragic end, Aphrodite honoured his memory by turning Melos into the first apple.

The fruit is also associated with the dead in Ireland. Samhain, the Gaelic festival of the dead, is also known as 'the feast of apples', as the fruits are left as offerings on graves and altars. Furthermore, coffins are commonly lined with apple wood, supposedly to grant the return of youth in the afterlife.

There was once a popular heirloom variety of apple—now extinct—to be found in America, which carries with it the perfect story of ghostly justice. The Micah Rood, or Bloody Heart, was 'sweet of flavour, fragrant, handsomely red outside, and while most of the flesh is white, there is at the core a red spot that represents human blood.'* The cultivar is supposed to have originated in the late 1700s in Franklin, Connecticut, on a farm owned by a man named Micah Rood.

Micah Rood was accused of murdering a travelling salesman, who had been found dead on his land under an apple tree, with a cracked skull and an empty purse. However, as no proof came to light that pointed to Rood's guilt, he was acquitted of the crime and allowed to go free. Later that year, however, the tree where the man had died began to bear red fruit with a bloody mark at the core, forever telling the world of his guilt. The farm soon fell into disrepair, and Micah Rood died destitute.

A similarly blood-marked variety, first recorded in 1883, is still thriving in Scotland. The Bloody Ploughman is named for a labourer who was shot dead for stealing apples from a Scottish estate to feed his family. Grieving and believing them cursed, his widow threw the apples into their yard rather than eat them; the next year, a tree sprouted there, growing apples with flesh that was as red as their skin.

Most famous of all apples is, perhaps, the one associated with the Garden of Eden and the origin of sin. The original tale of Adam and Eve in Genesis never actually names the fruit that grew on the Tree of Knowledge, and many scholars believe it more likely to have been a fig or pomegranate; but it was in medieval translations and paintings that it became an apple, likely as the fruit was more recognisable to the general populace. From thereon, the apple's good name was besmirched.

The medieval Church believed that enchanted apples, given of ill-will, could cause demonic possession, as it was through this sinful fruit 'that Satan continually rehearses the means by which he tempted Adam and Eve in the earthly Paradise'.† Jean Benedicti, a 16th Century Franciscan theologian, tells of Perrenette Pinay: a man who was found to be possessed by six devils after having eaten an apple and a piece of beef. Similarly, Jean-Francoir Fernel, a French physician, also tells of an unnamed man who became possessed after eating an apple.

*Charles Skinner; *Myths and Legends of Flowers, Trees, Fruits and Plants*
†Henri Boguet; *Discours Exécrable des Sorciers*

Around the same time as both of these accounts, a tale circulated around Europe about a supposed demon in Savoy in 1585. For a few hours, an apple lay at the edge of a busy bridge: otherwise inconspicuous, but for the 'great and confused noise' that it emitted, so much so that people were afraid to approach it. Crowds came to spectate, gathering round, and no one quite knew what to do with it until one brave man took a long stick and pushed it into the river. The tale ends, quite unremarkably, here, but it was later explained by Henri Boguet that: 'it cannot be doubted that this apple was full of devils, and that a witch had been foiled in giving it to someone.'*

GOLDEN APPLES

Golden apples have appeared in various legends throughout the ages. The golden apple of Eris, for instance, also known as the Apple of Discord, was used to set in motion the events that started the Trojan War. The incident was triggered by the marriage of Peleus and Thetis, for whom Zeus threw a great banquet. Eris was not invited due to her chaotic nature, and in retaliation she took a golden apple—upon it inscribed *Kallisti*, 'to the most beautiful'—and threw it into the celebration. Three goddesses, Hera, Athena, and Aphrodite, tried to claim it, and Zeus declared that Paris of Troy, who was known for being fair, would decide whom it belonged to. Each goddess tried to bribe him with skills or power, but Aphrodite offered him the love of the woman he already desired and who was already married to King Menelaus: Helen of Sparta. Paris chose Helen, and so the Trojan war began.

Eris's golden apple is said to have been stolen from Hera's garden, which was guarded by the *Hesperides,* the Daughters of the Evening, and a hundred-headed dragon. As with the fruit in the Garden of Eden, it has been suggested that Eris's creation wasn't actually an apple at all, but the fruit of the Argan tree, which closely resembles a small golden apple, and has an aroma similar to the baked fruit.†

Golden apples also play a role in Irish lore. These apples are supposed to grow in Emain Ablach (thought to be the Isle of Man or

*Henri Boguet; *Discours Exécrable des Sorciers*
†Michael Hübner; *Circumstantial Evidence for Atlantis in today's South Morocco*

the Isle of Arran), and are tended by Manannán mac Lir, an Irish sea deity and the guardian of Mag Mell, a part of the Otherworld reserved for those who died in glory. From here was cut the famous Silver Bough that appears in *The Voyage of Bran,* a silver branch laden with three perfect golden apples. When shaken, the apples would produce a music that would put anyone listening to sleep. If a man wished to enter the Otherworld before his appointed time, it was necessary to carry a token such as this to ensure his safe return.

ASPHODEL: Asphodelus spp.

> There asphodels are scattered through the night, like ghosts of young beseeching hands.
>
> William Faulkner, a note alongside drawings in a sketchbook

Most notable for their tall spike of white or yellow flowers, Asphodels are a common sight along the Mediterranean coastline. They grow particularly well in Corsica, where they are well-beloved as a national flower, and it is thanks to these that Corsicans have coined the phrase 'he has forgotten the asphodel', meaning that a man has been gone from his native land for so long that he must no longer remember it.*

For one brief year in 1648, yellow asphodel (*Asphodeline lutea)* was grown in the Oxford Botanic Garden as a point of curiosity, as John Parkinson, the apothecary and curator, wished to discover if the plant had any medicinal or edible purpose. His investigations turned up no useful information, and when interviewing locals in the Mediterranean was informed that the plant had 'no propertie appropriate unto it but knavery'. What knavery it was accused of was, sadly, not recorded.

In mythology, asphodel is best-known to the Greeks in the context of the famous Asphodel Meadows of the Underworld. After entering Tartarus, where the dead would be judged and then ordered based on the lives they had lived, the Meadows were where ordinary people—those who had neither committed great evil nor great good in their lives—were sent to live after death. Before entering they would drink from the river Lethe, which would cause them to lose any identity or memory of their life above. Then they would proceed to the Meadows: a ghostly replica of the living world, a land of complete neutrality, where they would continue to mechanically perform their daily tasks. Though the Meadows were not the torturous part of Tartarus (which was reserved for treacherous souls), they were not intended to be a peaceful place, either; the idea of spending eternity as part of a machine was meant to deter people from living a life of little impact, and ideally encouraged Greek civilians towards militarism rather than passivity. Those who did achieve greatness in their life would go to the Elysian Fields, where heroic souls would live in eternal contentment.

The Greek connection between asphodels and death likely comes from the plant's greyish leaves and pale flowers, which are waxy and yellow. Asphodel was planted on graves and devoted to Persephone, who often appears crowned with the blooms. Asphodels were said to be the favourite food of the dead, but were also eaten by the living poor, as the bulb could be baked, ground into flour, and then used to make bread.

*Dorothy Carrington; *The Dream-Hunters of Corsica*

Returning to Corsica, where stories of the supernatural thrive as fully as the asphodel, the plant plays an important role in the rites of the *mazzeri*, dream-hunters of the island.

These mazzeri (from *ammazza*, 'to kill'; and in other parts of the island known as *culpadori*, 'to strike a blow') are ordinary humans, but have been chosen to be supernatural messengers by the Qualcosa, the Corsican personification of destiny. Through their vocation they are able to be in two places at once; the living world, and the dreaming.

Whilst dreaming, they hunt at night (also leading to the name *nottambuli*, night walkers, or *sunnambuli*, sleep walkers) for wild boar and other local game. The weapon that they use for these dream hunts is called a *mazza*, and is a club cut from the root and stem of an asphodel plant, likely due to the plant's legendary connection to the Underworld. Once they have killed their prey, they then look into its face, and in it will recognise someone they know; usually someone within their village. In the morning they will recount what they have seen, and that very same person will know that they are fated to die within the year. However, if the hunted animal was wounded and not slain, that person will suffer an accident or illness instead.*

Though these divination skills are valued by Corsicans, the mazzeri themselves are often shunned from the villages they live in, and made to reside only on the outskirts of the community. This is because, though the mazzeru has no say in his vocation nor whose fate he might foresee, the Corsicans believe that the mazzeru does not hunt in his physical body, but as a spirit. When the hunter's spirit meets the spirit of someone in the village, who has assumed animal form in the dream world, his hunting of the animal severs the spirit from the body. Though the body can exist without its spirit for some time, eventually it will sicken and die.

There are tales of village mazzeri who would, once a year, form a militia to battle the mazzeri of a neighbouring community. These phantom battles were called *mandrache* and were performed in the dream realm; the mazzeri killed in the battle were condemned to die within a year, with some even found dead in their homes the very next morning.†

* Roccu Multedo; *Le Mazzerisme et le Folklore Magique de la Corse*

† Marie-Madeleine Rotily-Forcioli; *The Mazzeri I Have Known*

A related plant is the bog asphodel, *Narthecium ossifragum*. Unlike its sand-loving cousin, bog asphodel grows prolifically in the high, boggy moorlands of western Europe and the British Isles, and grows spikes of bright yellow flowers that have been used as a replacement for saffron.

Its Latin specific name means 'bone-breaker', as it was believed that cattle that grazed on it would develop brittle bones. It'smore likely that the bone conditions noted in sheep that also grazed in these areas were caused by a diet poor in calcium, and that the presence of asphodel was simply a coincidence thanks to the plant's preference of environment. However, it does cause in sheep a skin condition called *alveld,* or 'elf fire'. This is due to the animals grazing on parts of the plant that cause photosensitisation, which can create rashes and mild burns when the livestock that eat it are then exposed to sunlight.*

AUTUMN CROCUS: Colchicum autumnale

Thick new-born Violets a soft carpet spread
And clust'ring Lotus swelled the rising bed,
And sudden Hyacinths the turf bestrow,
And flowery Crocus made the mountains glow.

Homer, *The Iliad*

Despite its name and uncanny resemblance, the autumn crocus is not actually a true member of the crocus family. In fact, it is a highly toxic flower infamous for its involvement in the case of the last woman to be publicly hanged in the UK, in 1862. Branded 'the greatest criminal that ever lived' by the judge on her case, Catherine Wilson was only ever convicted of one murder, but it is believed that she was responsible for the deaths of at least six others. The nurse preyed on vulnerable and lonely people, such as her landlord (for whose murder she was finally arrested), and befriended them with the goal of being written

* J. T. Sibley; *The Way of the Wise: Traditional Norwegian Folk and Magic Medicine*

into their wills. Once she'd secured their inheritance, she gave them lethal doses of colchicine, and framed the deaths as suicides.

Colchicine's effects are similar to arsenic, and induce vomiting and cramps, followed by respiratory and cardiac failure. The Swiss physician Theophrastus (later known as Paracelsus) noted that Greek slaves would sometimes eat small amounts of the plant to become too ill to work.

It was named for the mythical Greek land of Colchis, where the witch Medea lived. The daughter of Hecate, Medea was a sorceress famous for her knowledge of poisons. Colchicum was one of her favourite plants to use against her enemies and grant to her favourites in exchange for youth and strength. It was colchicum that she gave to Jason, to help him yolk the fire-breathing oxen that guarded the Golden Fleece. Without Medea's knowledge, particularly that of plants, Jason would not have passed any of the legendary tests set to him.

Autumn crocuses flower in the latter half of the year, as their name suggests, and the flowers bloom long before the leaves appear, leading to the folk name *Mysterium.* Despite its mystery and toxic nature, it was believed in England that if it was seen blooming on a grave, it was a sure sign that the deceased were happy.*

* Hilderic Friend; *Folk-Medicine: A Chapter in the History of Culture*

AZALEA: Rhododendrum luteum

The numbers of bee-hives were indeed astonishing, and so were certain properties of the honey. The effect upon the soldiers who tasted the combs was, that they all went quite off their heads, and suffered from vomiting and diarrhoea, with a total inability to stand steady on their legs. A small dose produced a condition not unlike violent drunkenness, a large attack very like a fit of madness, and some dropped down, apparently at death's door. So they lay, hundreds of them, as if there had been a great defeat, a prey to the cruellest despondency. But the next day, none had died; and almost at the same hour of the day at which they had eaten they recovered their senses, and on the third or fourth day got on their legs again like convalescents after a severe course of medical treatment.

Xenophon, *Anabasis*,
an account of partakingof the honey of Trebizond on the Black Sea

Many are familiar with the sight of azaleas in public gardens, and may even grow one of these large, flowering shrubs themselves. But despite their popularity, toxins found in the plant can cause sickness and even death, as shown in the excerpt above. Regarding this tale, Pliny wrote in 77AD: 'What, in fact, could have been her [meaning Mother Nature] motive, except to render mankind a little more cautious and somewhat less greedy?'

This honey poisoning is caused by grayanotoxins that are found in the nectar of the rhododendron. Grayanotoxin poisoning is known as 'mad honey disease', and, though it is generally a wise idea to avoid it, contaminated honey is consumed in areas of Nepal and Turkey as a form of recreational drug.*

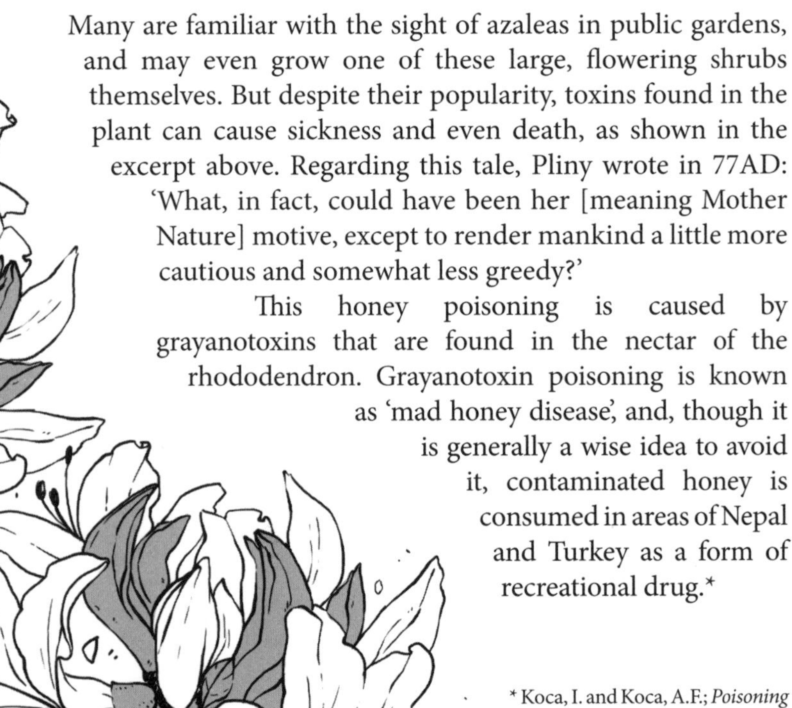

* Koca, I. and Koca, A.F.; *Poisoning by Mad Honey: A Brief Review*

Despite their danger, azaleas are readily eaten in China, where they are fried or served with beans, and the merchants who sell the flowers put them through a rigorous process of soaking and draining them twice daily to clean them of toxins. The azalea in China, particularly in the Yunnan province, is seen as having an 'arrogant and pure' energy; in the Naxi ethnic group, if a girl is called an azalea, it means that she is beautiful on the outside, but toxic within.* The Naxi people have another use for azalea leaves: the *dongba* priests burn the leaves inside homes to expel spirits or ghosts, and their churches are decorated with them for the same reason.

The Nu people of the same region tell of a strong woman named A-Rong, who made many clever contributions to her community. Learning of her skills, the head of the neighbouring village decided that he would force A-Rong to be his bride. Hearing of this plan, A-Rong ran into a forest of rhododendrons to hide; her suitor, deciding that no one else should have her if he could not, burned the forest—with her inside it—to the ground. For her bravery and intelligence, the Nu now honour A-Rong as a deity, and every year from March 15th to the 17th they hold a flower festival in her name, wearing and offering rhododendrons to her memory.

A story with similar themes is told by the Yi people in southwestern China. The Mayin Hua festival is held in February in honour of Miyulu, a young woman who was one of many to catch the eye of a particularly corrupt official. When he summoned her to his estate, she came wearing a white rhododendron flower in her hair, which she used to poison the wine that the both of them drank that evening. She lived just long enough to ensure that he died, before also succumbing to the poison. When her lover Zhaolieruo carried her body home, his grief was so great that his tears turned to blood, and forever stained the petals of the rhododendron red in her memory.

* Elizabeth Georgian and Eve Emshwiller; *Rhododendron Uses and Distribution of this Knowledge within Ethnic Groups in Northwest Yunnan Province*

BASIL: Ocimum basilicum

And so she pined, and so she died forlorn,
Imploring for her Basil to the last.
No heart was there in Florence but did mourn
In pity of her love, so overcast.

John Keats, *Isabella; or, The Pot of Basil*

A common pantry herb, basil is a familiar sight on many a kitchen windowsill. But in France it is believed to belong to the Devil, and will only grow if cursed at whilst planted; this has led to a French idiom, *semer le basilic*, or 'to sow the basil'—meaning to angrily rant. This belief comes originally from the Greeks, and then the Romans, who believed that the plant thrived best when hated.

However, in Italy and Romania it had somewhat more romantic connotations, and would either be given to a loved one

as a proposal, or displayed in the windows to show that the person living within was ready to receive suitors. *The Decameron,* a 12th Century collection written by Giovanni Boccaccio, tells of a tragic love story born from a pot of basil, which was later adapted by the poet John Keats.

The story tells of Lisabetta—later called Isabella by Keats—who lived in Messina with her three wealthy merchant brothers. Though her brothers expected her to marry well, she fell in love with Lorenzo, a poor man who worked for her family. When the brothers found out, they murdered Lorenzo and buried his body, then told Lisabetta that he had been sent abroad on business.

The more time that passed without word from Lorenzo, the more Lisabetta despaired. Each night she called out to him, begging him to return, and one night his ghost did just that, telling her what had happened and where he had been buried. The next day she slipped away to find his body, but was unable to carry it alone. Instead she cut off the head and, to keep it close and safe, buried it in a garden pot which she planted over with basil.

Every day she wept over the plant, watering it with her tears. The plant flourished, but she grew weaker from grief, until her brothers—discovering the reason for her great sadness—took away the pot and its gruesome secret. Without Lorenzo or her beloved basil plant, Lisabetta died not long after.

A relative to the common kitchen basil is Tulsi (*Ocimum sanctum*), also known as holy basil. A sacred plant in India, holy basil is dedicated to the Hindu god Vishnu. A legend explains the connection: it is an incarnation of the goddess Tulsi who, when she was still a mortal woman, threw herself onto her husband's funeral pyre in grief. At this point, her soul was transferred into the plant.

Another legend claims that holy basil is actually an incarnation of Lakshmi, Vishnu's wife. Vishnu is said to feel pain if the plant is damaged, and will refuse to hear the prayers of anyone who mistreats it. However, placing a leaf of holy basil on a body after death will ensure that Vishnu sees the soul of the departed, and welcomes them into heaven.

BITTERSWEET: Solanum dulcamara

> I should not think it strange, for 'tis a physic
> That's bitter to sweet end.
>
> William Shakespeare, *Measure for Measure*

Also known as woody nightshade, bittersweet is a direct relative of deadly nightshade that is surprisingly common, and easily overlooked. This prolific climber grows well in hedges and amongst brambles, and is easily mistaken for redcurrants; much to the regret, no doubt, of any who eat it.

This vine rambles freely through woodlands and marshy areas, and can be festooned with tiny purple and yellow flowers even at the same time that it bears berries. These little berries start off green, turn yellow, and finally red when ripe. The name *dulcamara* comes from *dulce amara*, or 'sweet-bitter', referring to the flavour of these berries: they taste bitter at first, and then sweet afterwards. The name was given by English botanist John Gerard, who translated the words the wrong way around, but the epithet stuck nonetheless.

This sweet taste, and the sickly-sweet scent it gives off when bruised, is a result of dulcamarine. Twinned with solamine, another compound found in the plant, an overdose of these berries can lead to paralysis of the central nervous system, respiratory collapse, and convulsive death. A non-lethal dose causes a temporary loss of speech, a symptom which was once believed to be the result of a witch's curse.

Despite its toxicity, bittersweet is thought across the British Isles to have protective powers against magic and witchcraft. Shepherd-lore tells how the wood of the bittersweet vine can protect sheep and pigs against the evil eye, and hide them from witches. Similarly, John Aubrey recommended it as 'a receipt to cure a horse of being hag-ridden: take bittersweet and holly. Twist them together, and hang it

about the horse's neck like a garland; it will certainly cure him.' Any horse left in the field at night was in danger of becoming hag-ridden—witches might mount a horse while it slept and ride it through the countryside, leading to exhaustion and injury, and leaving the beast useless to its owner the next day.

In Norway, a combination of bittersweet, heath spotted orchid, and tree sap would be smeared on both people and animals to protect them against demonic influences.* Germanic folklore also ascribes to it an ability to protect against demons, as well as connecting it to elves and fairies, in light of which it has the name *alprauke,* or elfwort.† Many plants with a climbing habit, such as columbine and honeysuckle, are given the same name.

* Reimund Kvideland and Henning Sehmsdorf; *Scandinavian Folk Belief and Legend*
† Johann Friedrich Karl Grimm; *Remarks of a Traveller Through Germany, France, England and Holland: In Letters to his Friends*

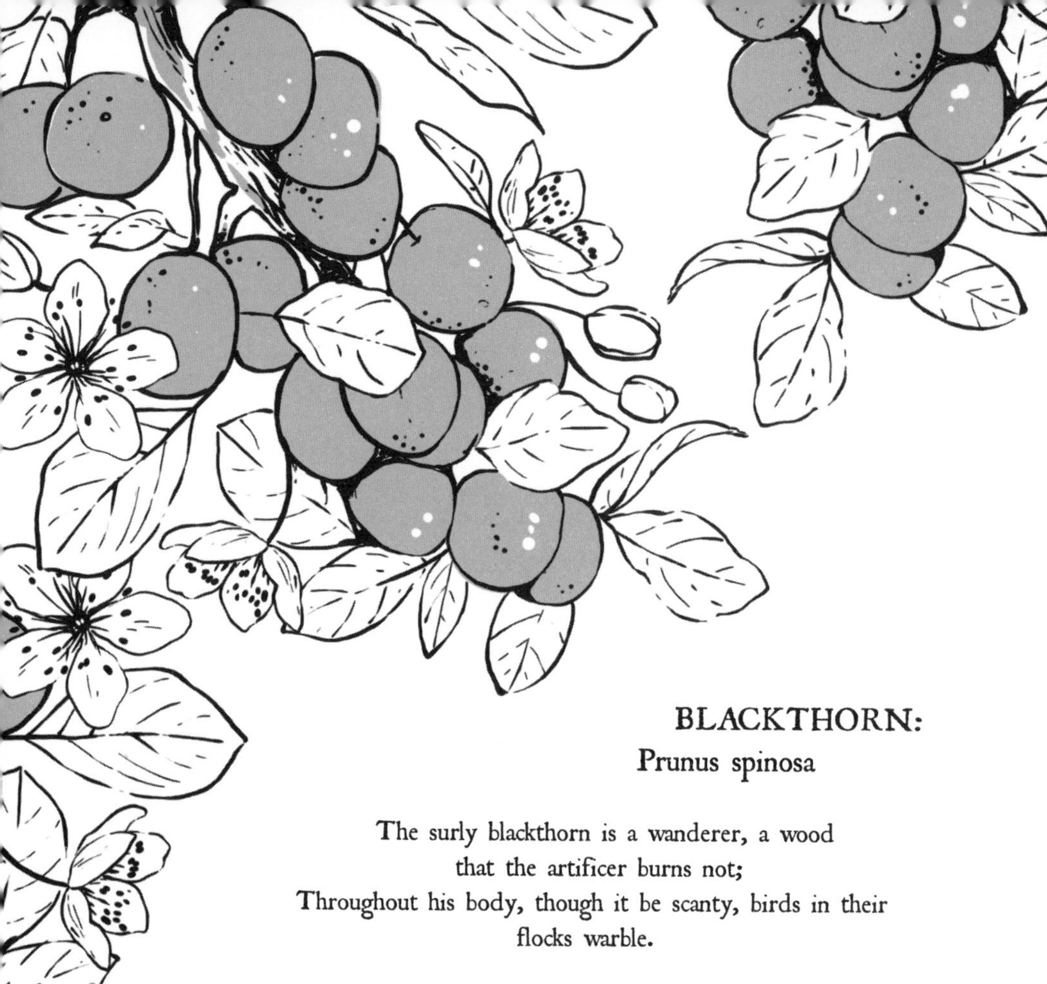

BLACKTHORN:
Prunus spinosa

The surly blackthorn is a wanderer, a wood
that the artificer burns not;
Throughout his body, though it be scanty, birds in their
flocks warble.

Unknown author, *The Violent Death of Fergus mac Léti*

For anyone familiar with picking sloes to flavour gin, the blackthorn is no doubt already associated in their mind with autumn and the darker side of the year. And they won't be alone; this long-thorned shrub has been associated with winter and darkness for centuries.

It is one of the first hedgerow shrubs to bloom in spring, usually bearing white flowers from March to mid-April. However, should it bloom too early it begins what is known as a 'blackthorn winter'—a warm period at the end of March, and a cold snap in April. Later in the year, the number of berries borne by the branches foretells of the coming winter. An increased number of sloes indicates that the winter will be cold and long:

Many sloes, many cold toes.

Michael Denham, *The Denham Tracts*, 1846–1859

Much of the lore of the blackthorn comes from the British Isles and northern Europe. Most commonly it is presented as the converse to the benign whitethorn, known more commonly as the hawthorn (*Crataegus monogyna*), which it often grows close to. Of these two sister trees, Hawthorn was seen as the more dominant.

Where whitethorn was deemed 'Queen of the Woods' and seen to be lucky, it is the blackthorn that has had its reputation smeared. Unlucky and cursed in every way, it was said to grow out of the bodies of heathens, whereas from the bodies of Christians would grow the blessed whitethorn.* On the morning of May Day (usually May 1st), when the maypole would be crowned with white- and blackthorn alike, it was a 19th Century custom to hang garlands over the doors of local girls, each one conveying an explicit message about the village's opinion of those who lived inside. Whitethorn was the most complimentary, but blackthorn was reserved for those believed to be shrewish in nature.† The worst insult that could be given was to hang a bunch of nettles over the door.

The origin of this unfortunate reputation may come from the dangerous nature of the sloe's thorns. Though the hawthorn also bears spines along its branches, the thorns of the sloe are particularly long and hard, and prone to causing severe bleeding. Though the tree itself is not toxic, the bark is covered in bacteria that cause inflammation and infection that can lead to septicaemia. These thorns were believed to belong to the Devil, and were used by him to mark the skin of witches.‡ These marks were used as evidence that an accused woman was a witch, though in reality the victim may have simply displayed a birthmark, an insect bite, a scar, or a wart. Many women were condemned on no other evidence than this mark; in fact, Anne Boleyn, the second wife of Henry VIII, had just such a birthmark on the back of her neck, which contributed to public vitriol against her.

* William Thiselton-Dyer; *The Flora of Middlesex*

† Katharine Tynan and Frances Maitland; *The Book of Flowers*

‡ Fred Hageneder; *The Meaning of Trees*

The thorns of the wood were said to have divinatory powers, and in Wales were used to test the faithfulness of a lover. In Llanblethian in Glamorgan, thorns would be dropped into the village well by young maidens: if the thorns floated their lover was faithful, but if they sank, the truth was rather more doubtful. However, if the thorns whirled around then they showed them to be of a cheerful disposition, and if they sank just a little, they were likely to be of a stubborn nature!

The spines would also be used by English witches, in sticking into wax images or placing beneath the saddles of horses to throw the rider, and were often called 'pins of slumber'. Blackthorn wood was also ideal for creating 'blasting' wands with which to cast curses. So associated with witches did the blackthorn become in the 16th Century that the shrub was denounced by the Church as a witch's tool, and the wood of it used for the pyres they burned them upon.

It wasn't just witches that the tree became symbolic of. It is depicted throughout many European fairy tales as a general tree of ill omen, and paid particular attention to by the Scottish and Irish Celts. They called it *straif* (thought to be the origin of the English word 'strife'), and believed it to be a keeper of all dark secrets. It was associated with the Cailleach, the goddess of winter, an old woman who would emerge at Samhain (November 1st) to take over the year from the summer goddess Brighid. In Scotland she is known as Beira, the Queen of Winter. Wearing a blue veil and with a raven on one shoulder, she would carry a blackthorn staff with which she brought about storms and rough weather. With its dark wood and twisted form, the blackthorn isn't dissimilar in appearance to its patron, with both the tree and the goddess being known as the 'dark crone of the woods'.

Though partnered with Cailleach, the Irish Celts believed that the tree was also inhabited by the *Lunantisidhe*: unfriendly moon fairies who would use their long arms and fingers to climb the branches of the blackthorn and curse those who came near it. The only time they left the tree was at the full moon to worship Arianrhod, the goddess of the moon. This was the safest time to gather sloes or to cut the wood to make into shillelaghs.

The shillelagh, also known as a *bata,* is used in traditional Irish stick fighting. Though it can be made from other woods such as holly, oak, or ash, blackthorn is preferred as it is a hard wood, and the knot of roots at the base can be shaped easily into a rounded knob for the top of the stick. This blackwood staff was said to protect a person whilst walking in haunted or cursed places; but it wasn't long before those who carried it were suspected of practicing witchcraft, such as in the case of Major Thomas Weir, who was burned in Edinburgh in 1670 alongside the blackthorn staff that was said to be the chief instrument of his sorceries.

It is interesting that it is the blackthorn that has gathered about itself a darker reputation, whereas its sister tree, the hawthorn, has remained untarnished. Trimethylamine, which occurs in decaying flesh, also naturally occurs in the flowers of the hawthorn, which gives the tree an unpleasant rotting odour towards the end of the flowering season. Perhaps for this reason, the Welsh believe that it is a gateway to *Annwn*, the Welsh Otherworld and land of the dead.

BLUEBELL: Hyacinthoides non-scripta

He laid him down,
Where purple Heath profusely strown,
And Throat-wort, with its azure bell,
And Moss, and Thyme, his cushion swell.

Sir Walter Scott, *Rokeby*

In the British Isles, bluebells are seen as an ancient herald of spring, and are closely associated with the fairy world thanks in no small part to their delicate, bell-like flowers—an association shared by other plants of a similar shape. Patches of bluebells are said to be full of magic, and are to be feared as much as admired. The ringing of the bells was said to summon fairies to their gatherings, but any mortal who heard the sound would be fated to die before the meeting took place.

In Finland, the bells exist for the benefit of mice, not fairies. Called *Kissankello,* the cat's bell, the myth tells of a group of mice that were plagued by the attention of a particular cat. They had on them a bell that they plotted to tie around the cat's neck, but none of them were brave enough to do the deed. As they were arguing over who would take it, a fairy overheard and offered to buy the bell from them; she then took it and turned it into a blue flower, which would ring whenever the cat came near.

The Romans called it Venus's Looking Glass, in honour of a mirror that Venus possessed which would reflect whatever it faced as a more beautiful version of itself. After the goddess dropped it in a field, a shepherd picked it up and fell in love with his own beauty. Cupid, seeing this and fearing for the consequences, shattered the mirror into glittering fragments, which were then turned into bluebells.

Despite its beauty, bluebell bulbs are extremely toxic. They contain scillaren, a glycoside similar to the chemical found in foxgloves, another bell-shaped flower favoured by the fair folk. Exposure to scillaren lowers the pulse and causes irregularities in the heartbeat, which is perhaps why it was believed that wandering into a ring of bluebells would put a man into an endless enchanted sleep.

Before 1934, the scientific name for the bluebell was *Endymion non-scripta,* in honour of Endymion, who was the lover of Selene, the Greek goddess of the moon. Taken by his beauty and unwilling to share it with anyone else, she put him into an eternal sleep so that she alone could gaze upon him. Another sleep-related tale comes from Ireland, where they tell of Gráinne, a princess who was instructed to marry Fionn mac Cumhaill, a mythical hunter. However, she was in love with the demigod

Diarmuid, and at her wedding mixed the juice from bluebell bulbs into the wine of those present, so that as they slept she could elope with Diarmuid.

Though bluebells have historically been recommended by herbalists to prevent against nightmares, there is no proof that bluebells do have any soporific properties, and the poisonous nature of the plant suggests that if it is to bring any sleep at all, it is of the more permanent kind.

However, the gummy sap of the bulbs does have another application: it can be purposed as a glue that was once used to bind fletchings to arrows, and was also used as a glue for bookbinding, as the insecticide nature of the sap would deter insects from damaging the pages.

BROAD BEANS: Vicia faba

And so soon he began to crawl,
then stood firm as a mountain.
When he began to feed himself,
he planted broad-beans aplenty,
broad and wind-fluttered beans,
and lush grain ripening in rows,
wheat and hemp thick and rich,
and melons sprawled everywhere.

Chinese folk poem, *Birth to Our People*

As a popular crop food across the world, broad beans aren't even remotely poisonous. But their place in this book is earned through their curious connection to the dead that started with the Roman Empire, and continues in our customs even today.

To understand why beans were of such note to the Romans, it's important to understand their views on the dead in the particular period when this tradition began. The dead were held as something to be appeased and honoured, for if the living fell out of their good graces, they would return to haunt them as the restless dead, called *lemures,* or *larvae.* These were grotesque and twisted ghosts that would torment their living relatives, and bring misfortune, madness, or disease to the affected households. So common was this fear that a widely-used term for mortal madness was *larvaetus,* or 'possessed by the dead'. The recently deceased were more likely to return if they had died in a difficult fashion, such as those who had died violently (except during military service), by suicide, or those who had no tomb or resting place.

At the beginning of May, the festival of Lemuralia was held, a period where rituals would be performed to ward off these ghosts. The head of the household would rise at midnight, and walk the house whilst throwing beans over their shoulder, reciting: 'I send these; with these beans I redeem me and mine'. Incidentally, this led to the entire month of May being viewed as bad luck for weddings, leading to the proverb *Mense Maio malae nubunt:* 'Bad girls wed in May'!

So what was the connection between beans and the hauntings of *lemures?* Romans believed that the spirits of the dead travelled up the hollow stems of broad beans, and stayed there until they matured into the fully-grown beans. A field full of beans, to someone who believed that every bean on the stalk was a restless spirit, must have seemed a fearful thing. There was no fear of eating these beans, however: anyone who has eaten enough of them knows that they can cause chronic gas, which was seen as a sign that the dead were escaping to where they were meant to be! Beans were also left out in the home as a protective measure, with the hope that the *lemures* would take the spirits in the beans instead of those that belonged to the still-living.

The fear of beans was no laughing matter for Pythagoras, the famous Greek philosopher. There are multiple accounts of how he died; however, one legend describes how, when he was being pursued, he found himself trapped in a field of beans. Rather than step on the plants, he hesitated long enough for his murderers to catch up with him, who then beat him to death.

The association of beans with the dead has not remained bound to history. In modern-day Rome, bean-shaped cookies called *fave dei morti* ('beans of the dead') are still baked and eaten on the Italian Day of the Dead, November 2nd.

The custom travelled as far as the British Isles, where—in Yorkshire in particular—the dead were believed to reside in the flowers of the bean. It was generally observed that more accidents, specifically in coal pits, were likely to occur during the period in which beans flower.* These accidents were more likely caused by heavy spring rains softening the ground around the mines, but the association has stuck fast.

* Sidney Oldall Addy; *Household Tales with Other Traditional Remains*

BRUGMANSIA: Brugmansia suaveolens

Relentless Time! that steals with silent tread,
Shall tear away the trophies of the dead.
Fame, on the pyramid's aspiring top,
With sighs shall her recording trumpet drop;
The feeble characters of Glory's hand
Shall perish, like the tracks upon the sand;
But not with these expire the sacred flame
Of virtue, or the good man's awful name.

William Lisle Bowles, *The Grave of Howard*

Brugmansia is a large tree or shrub native to tropical regions of South America. The common name angel's trumpet relates to the trumpet-shape of the large, white flowers, which hang downwards from the branches and can grow up to twenty inches in width. The name angel's trumpet is also occasionally used for the Datura genus; a genus that the brugmansia once belonged to, before being reclassified in 1805. Though they are popular ornamental plants and are widely cultivated, the species is listed as extinct in the wild; it is thought that whatever animal previously dispersed its seeds has also become extinct, making it now entirely dependent on humans for survival.

It was once called *Floripondio*, an old term dating back to the Spanish Conquistadors. Bernabé Cobo in 1653 described them as follows:

This flower is the biggest of all those produced by trees and bushes, is beautiful to look at and is white; it is a palm in length and five points bend back from the very wide mouth ... Their fragrance is so strong and intense that they must be sniffed from afar rather than from near at hand, and only one of these flowers in a room perfumes it so much as to be irritating and usually produces a headache.

The strong scent of the plant that Cobo noted is a particular calling card of the brugmansia. The pollen alone is said to provoke vibrant

dreams, and simply being in the same room as a flowering plant can have an effect. In South America, where it originated, there is a belief that sleeping underneath the flowers could cause insanity.*

It is not just the pollen that is hallucinogenic; all parts of the plant are highly poisonous. Poisoning from the brugmansia stimulates, but then depresses, the central nervous system, causing hallucinations, delirium, incoherent speech and convulsions, followed by unconsciousness. Its ability to induce comas was made use of by the Chibchas of pre-conquest Bogotá in Columbia: when a chieftain or warrior died, they would mix the plant with maize beer and tobacco leaves at the funeral, and give it to the wives and slaves of the deceased to induce a stupor before being buried alive with their masters and husbands.†

Despite its dangers, it has been, and still is, a popular hallucinogenic drug. In the Victorian era it was grown as a house plant, intended to have tea served beneath it; the drinker would then tap the flower to transfer the pollen into their mug. The resulting high was remarked to be similar to taking LSD.

However, taken in too large a dose or if the plant is ingested directly, the hallucinogenic effects have been described as 'terrifying

* T. E. Lockwood; *The Ethnobotany of Brugmansia*
† Richard Evans Schultes; *The Plant Kingdom and Hallucinogens Part III*

rather than pleasurable',* and can induce a powerful trance characterised by a loss of awareness that the person is hallucinating, as well as a psychosis that can cause visualisations of being something other than human, such as an alien or demon. In Germany in 2003 a teenager, after drinking one cup of tea made from the leaves, suffered these visions to such an extent that he amputated his own tongue and genitals. This tea is made in Madeira for recreational use, but is known as the Devil's Tea due to its dangers.

Horrific though this potential might seem, brugmansia has been used in ritual and traditional medicine by the Andean tribes of Peru and the Amazonian tribes of Ecuador for thousands of years. Humans have used hallucinogenic plants as an intermediary between the human and spiritual world since ancient times, and brugmansia, known to the Andeans as *mishas*, plays a pivotal role in rituals of coming of age, conversations with ancestor-spirits, and divination.

It was thought that the spirit of the brugmansia manifested as a bull, and that by tying the leaves in the shape of a cross on the forehead, you would be gifted the ability to see if a man's heart was good or evil. A smaller variety, *B. insignis*, could be used in the same way to enhance dreaming. This species manifested itself in the image of a hunting dog, and would help to find lost things in dreams. Other varieties were said to manifest as a puma, a bear, or a snake.† And brugmansia could manifest itself as more than just animals. Unruly children would be exposed to this plant as a correctional measure, as it was believed that it would summon the spirits of their ancestors to speak to and admonish them for their bad behaviour.

* Christina Pratt; *An Encyclopedia of Shamanism*

† Vincenzo De Feo; *The Ritual Use of Brugmansia Species in Traditional Andean Medicine in Northern Peru*

BRYONY, WHITE: Bryonia dioica

... they take likewise the roots of mandrake, or the roots of briony, which simple folk take for the true mandrake, and make thereof an ugly image, by which they represent the person on whom they intend to exercise their witchcraft...

William Coles, *The Art of Simpling*

A part of the cucumber family, white bryony is an enthusiastic climbing plant that appears without warning and can rapidly take over a hedgerow. Some of its folk names, such as dead creepers and death warrant, are unrepentant about its deadly nature; its berries are particularly poisonous, and just ten of them are enough to kill a child.

It is worth noting the existence of a similarly-named plant, black bryony. Though botanically unrelated, the two plants are remarkably alike in appearance. They both bear small flowers in summer and red berries in winter, but the black bryony's leaves are large and heart-shaped, whereas the white has lobed leaves and tendrils to help it climb. Both are worth giving a wide berth; they are equally as poisonous as the other.

White bryony is most famous for being sold as counterfeit mandrake in the British Isles. As mandrake takes at least three years to mature, plants with large, tuberous roots such as bryony, cuckoo pint, and enchanter's nightshade would be sold as 'English mandrake' and attributed with

the same powers as true mandrake. Due to the reputation of mandrake for its figure-like roots, the plants would be dug up and the roots carved into a rough human figure, then buried again until the fresh cuts scarred over.

Another technique, for a different purpose, was described by Sir Thomas Browne in 1646. People would 'carve out the figures of men and women, first sticking therein the grains of barley or millet where they intend the hair to grow; then bury them in sand until the grains shoot forth their roots which, at the longest, will happen in twenty days; they afterwards clip and trim these tender strings in the fashion of beards and other hairy parts...' These hairy figures weren't to trick people into buying them as mandrake, but instead were to be entered into 'Venus Night' competitions, where awards were given for the most accurately-carved forms!

Bryony root was most favoured for carving, due to the size to which it can grow. It can reach up to 30cm long and 10cm wide, and the botanist Nicholas Culpeper wrote in 1663 that he had been shown 'a root weighing half a hundredweight [56lbs], and the size of a one-year-old child'.

BUTTERCUP: Ranunculus spp.

Ranunculus, who with melodious strains
Once charmed the ravished nymphs on Libyan plains,
Now boasts through verdant fields his rich attire,
Whose love-sick look betrays a secret fire;
Himself his song beguiled and seized his mind
With pleasing flames for other hearts designed.

René Rapin, *Of Flowers*

The name ranunculus may conjure images of delicately spiralled petals and perfectly spherical flowers, but the wild buttercup is vastly different in appearance to its cultivated cousin. Most common in our meadows and gardens are *R. repens* and *R. acris*, the creeping and meadow buttercups, cheerful little weeds that spread voraciously and brighten fields and roadsides with their yellow flowers. The name ranunculus comes from their disposition towards growing near rivers and streams, *rana* and *unculus* meaning 'little frog', as they are as plentiful as frogs in spring. They are also called crowfoot or crowflower in some areas, due to the shape of the leaves.

Cheerful prophets of spring though they might be, the leaves contain the chemical ranunculin, which can cause dermatitis in humans when handled, and severe blistering of the mouth if eaten. This effect was made use of by beggars in the 1500s, as recorded by botanist John Gerard, who called it a 'furious biting herb': 'cunning beggars do use it to stampe the leaves, and lay it unto their legs and arms, which causeth such filthy ulcers as we day by day see (among such wicked vagabonds) to move people the more to pitie'.

A Persian legend tells of how the flower came to be. Before buttercups grew on the earth, there was once a young prince who favoured wearing green and gold. He fell in love with one of the beautiful nymphs who lived near the palace, and declared his love by singing to her through night and day in the hope that she would return his affections. Here the story has two possible endings—either the nymph spurned his affections, and when he died of heartbreak his body turned into a ranunculus; or the nymph grew so sick of his singing that she silenced him by turning him into the flower herself!

CARNIVOROUS PLANTS

Then, methought, the air grew denser, perfumed from an unseen censer
Swung by Seraphim whose foot-falls tinkled on the tufted floor.
"Wretch," I cried, "thy God hath lent thee—by these angels he hath sent thee
Respite—respite and nepenthe from thy memories of Lenore;
Quaff, oh quaff this kind nepenthe and forget this lost Lenore!"
Quoth the Raven, "Nevermore."

Edgar Allan Poe, *The Raven*

Most of the plants we know best draw nutrients from the earth and water around them. In comparison to this harmless existence, carnivorous plants—those that derive some, or most, of their nutrients from the trapping and consumption of animals and insects—have always been an anomaly that fascinates us. This characteristic has mostly evolved from necessity caused by nutrient-poor environments, such as marshes and rocky alpine land, where the soil is thin and not ideal for healthy growth.

As set out by Charles Darwin in 1875 in his *Insectivorous Plants,* there are six basic types of carnivorous plant: pitfall traps, also known as pitcher plants; adhesive traps (such as the sundew); snap traps, popularised by the venus fly trap; suction traps; lobster traps; and pigeon traps. All of these have evolved various 'hunting' mechanisms to trick, ensnare, and otherwise digest living creatures to ensure their own survival.

PITCHER PLANT: Nepenthes spp. and Sarraceniaceae spp.

The 'pitcher' part of a pitcher plant is a specially modified leaf that is curled into a watertight, cup-shaped trap to hold digestive liquids. Often, the edge of the pitcher is lined with nectar to attract its prey, but the walls, once touched, are slippery and prevent climbing back out. The liquid inside of the pitcher then drowns and gradually digests the prey. Some of the largest species, such as *Nepenthes rafflesiana*, are even able to catch and digest bats, lizards, and rats.

The genus *Nepenthes* is named for the fictional Greek medicine that was meant to cure sorrow. In the Odyssey, nepenthe was given to Helen of Troy by Paris to help her forget her old home. Despite the genus of pitchers named for it, it was thought by ancient writers such as Pliny and Dioscorides to be the herb borage, but suspected by more contemporary academics to be opium.

One of the most remarkable-looking species is the *Nepenthes naga,* a pitcher plant endemic to the Barisan Mountains of Sumatra. Local folklore claims that dragons once lived in this area, and the Naga certainly lives up to the stories; *naga* is the Indonesian word for 'dragon', and the appendage that grows on the underside of its lid is shaped much like a snake's (or dragon's) tongue.

Another species of pitcher plant is the *Heliamphora*, the marsh pitcher plant. The particle *heli* in the name was once mistakenly thought to come from the Greek *helios*, meaning sun, leading to the name sun pitcher which, though incorrect, has stuck. The correct origin is *helos*, meaning marsh.

Endemic to South America, there exists a cultivar named the patasola, after the mythical Patasola vampire. This beautiful woman appears in jungles and mountain ranges late at night, luring lone herdsmen, loggers, or miners into the undergrowth. When they are sufficiently lost, she transforms into her true form—a one-legged, cloven-footed creature with great teeth and large eyes—and devours them. When she is sated, she climbs a tree and sings the following to herself:

I'm more than a siren,
I live alone in the world:
and no one can resist me
because I am the Patasola.
On the road, at home,
on the mountain and the river,
in the air and in the clouds,
all that exists is mine.*

SUNDEW: Drosera spp.

The sundew genus comprises almost 200 species of plant that can be found across a vast range of climates, from Alaska to New Zealand. However, the majority of these species seem to grow in hotter parts of the southern hemisphere, including Australia, South America, and southern Africa. The scientific name *Drosera* comes from the Greek *drosos*, meaning dew, and refers to the sticky drops on the plant's fringes.

The plant is characterised by these fringes, or tentacles, which cover its surface. At the end of each tentacle is a single drop of sugary, sticky liquid, which attracts prey and then stops it from escaping. These tentacles then curl inwards to bring the prey into contact with as many of them as possible, speeding up ingestion. The prey either dies of exhaustion from struggling, or is digested by the enzymes in the secretions.

Though dangerous to wandering insects, sundews have proven useful to humans in multiple ways. They have been used as a medicinal herb in Italy and Germany since the 12th Century—named *herba sole,* they were administered in tea or preparations for coughs and asthma. The corms are also considered an edible delicacy by some Australian Aborigines, or used to dye textiles; in the Scottish Highlands, *D. rotundifolia* is used to prepare a purple or yellow dye.†

D. rotundifolia, or the Italian Sundew, is also used to produce a liqueur called Rosolio (after the Latin *ros solis,* 'dew of the sun'), which is produced using a recipe that has barely changed from its original 14th Century version. This bright yellow spirit originated in Italy, and

* Javier Ocampo Lopez; *Mitos, Leyendas y Relatos Colombianos*
† Edward Dwelley; *Dwelley's Illustrated Scottish-Gaelic Dictionary*

was initially taken as a medicine and aphrodisiac. This was common of many early cordials, which were prescribed as alcoholic medicines to revive the heart and soul, before gaining a wider popularity as a drink taken for pleasure.

Early versions of Rosolio were made from the petals of the sundew, together with sugar, vanilla, water, and raw spirits. It is also thought that galingale and grains of paradise—both members of the ginger family—were included to add warmth. Another version of the recipe from the 1600s, comes from Sir Hugh Platt's *Delightes for Ladies:*

Take of the herb Rosa-Solis, one gallon, pick out all the black moats from the leaves, dates half a pound, Cinnamon, Ginger, Cloves of each one ounce, grains half an ounce, fine sugar a pound and a half, red rose leaves, green or dried four handfuls, steep all these in a gallon of good Aqua Composita in a glass close stopped with wax, during twenty days, shake it well together once every two days. Your sugar must be powdered, your spices bruised only or grossly beaten, your dates cut in long slices with the stones taken away. If you add two or three grains of Ambergris, and as much musk in your glass amongst the rest of the Ingredients, it will have a pleasant smell. Some add the gum amber with coral and pearl finely powdered, and fine leaf gold.

It was noted by botanist Susan Verhoek-Williams that sundew leaves were used by French sorcerers in the middle ages to ensure that they did not tire in their work. The plants were said to glow at night, and were favoured by woodpeckers, which would harden their bills upon it. These attributes are also given to the moonwort fern (*Botrychium lunaria*), an occurrence which is common in early manuscripts; many have been subject to mistranslations, misidentifications, poor illustrations, or wild guesses at the identity of plants which have since been reclassified under different names. As a result, some tales and superstitions said to belong to one genus may also be claimed to apply to another as well.

BUTTERWORT: Pinguicula spp.

Butterworts are a genus of carnivorous plants that grow across Europe, North America, and northern Asia. Similar to the sundews, the flat leaves are coated with a sticky enzyme that attracts and traps their prey, which curl inwards to prevent it from escaping. They grow best in damp, nutrient-poor sites such as bogs and wet heaths, and are also known as bog violets. The name *Pinguicula* was given to them by Linnaeus, which translates to 'little greasy one'.

The plant is particularly loved by the Scottish, who would hang them over the doors of houses where corpses awaited burial, in order to stop the dead from rising again. It was also believed that if a woman should pluck nine roots of the butterwort, and knot them together into a ring, she could place it in her mouth and kiss a man to ensure that he would forever be obedient to her.

The name 'butterwort' comes from another Scottish and Hebridean belief that the plant could protect a cow's milk from being turned by witches. The leaves could be rubbed on a cow's udders to protect it from evil influences, a protection that could also be gained by the cow eating the plant itself. It is thought that butterwort may be the identity of the *mothan,* or *moan* in Ireland, a mythical herb said to be a protective charm for cattle against witchcraft. In the Hebrides, a man who has miraculously escaped death is said to have 'drunk of the milk of a cow that ate the mothan'.*

* Alasdair Alpin MacGregor; *The Peat-Fire Flame: Folk-Tales and Traditions of the Highlands and Islands.*

CHERRY LAUREL:
Prunus laurocerasus

You may see the laurel's girth,
big of bloom, give fragrant birth
to the oread whose hair,
musk and darkness, light and air,
fills the hush with wonder
there.

Madison Julius Cawein, *The Land of Hearts Made Whole*

A favourite of Victorian and modern gardeners alike, the cherry laurel is the most common plant known simply by the name 'laurel', and in America is regularly known as English laurel. This fast-growing hedging plant has glossy, dark-green leaves and is cultivated in temperate regions worldwide.

Popular though it is, the plant—both the leaves and the fruit pips—are a reliable source of cyanide, which starves the nervous system of oxygen and can lead to death. Though there have been no reports of anyone dying from exposure to the plant, unwary gardeners have reported feeling light-headed and dizzy whilst transporting laurel cuttings in a vehicle. Not ideal for anyone in control of machinery, but certainly an effect that Edwardian insect collectors took full advantage of: killing a specimen without damaging its delicate body was a challenge to most, so a popular technique involved capturing the insect in an airtight jar that contained crushed laurel leaves, and letting the cyanide fumes do the job.

A distillation of the leaves can also be a source of hydrocyanic acid, also known as prussic acid. This cherry laurel water was favoured by the Roman emperor Nero to poison the wells of enemy cities, and was instrumental in the murder of Sir Theodosius Boughton by his brother-in-law in 1780, one of the best recorded incidents of cyanide poisoning at the time. Since Captain John Donellan, the accused, stood to inherit quite a fortune should Boughton die before his 21st birthday, he conspired to poison Boughton with laurel water. Despite an initial verdict of death by illness, and Donellan's many attempts to thwart the later investigation, Donellan was finally exposed by the victim's mother, who identified the bitter almond smell associated with cyanide as belonging to the drink that had been given to Boughton on the day of his death.

CORN COCKLE: Agrostemma githago

What hurt it doth among corne, the spoil of bread, as well in colour, taste and unwholesomeness, is better known than desired.

John Gerard, *Great Herball*

An attractive, delicate wildflower, corn cockle is now increasingly rare, but used to grow rampant across the Northern Hemisphere. Once commonly seen alongside poppies, cornflowers, and other wildflowers that thrive in the ploughed earth of agricultural land, proof of corn cockle's hardy nature has been found in Stone Age villages and even in the area around Pompeii dating from before the city was destroyed.

The decline of corn cockle, in crop fields at least, is largely due to a determined campaign by farmers to reduce the chance of the plant finding its way into harvests. The whole plant, but particularly the seeds and oil from such, contain the glycoside githagin and agrostemnic acid, which turn flour rancid and produce a bread that is grey and bitter. Just a pinch of the seeds is enough to kill a horse or human being, as it paralyses the respiratory system until the victim dies of suffocation. Most modern appearances of the plant have been swiftly dealt with through the use of pesticides, however in France an integral part of the *Fête des Brandons* on the first Sunday of Lent is the pulling of cockle from corn harvests, and in England corn would be similarly treated during events known as corn showings.*

In Lithuania, a snake-like creature called the *kūkalis* (literally, 'corn-cockle') makes a nuisance of itself by blighting crops of corn. The *kūkalis* is closely related to the kaukas, a harvest hobgoblin, which performs the same role.†

CUCKOO PINT: Arum maculatum

And mother, find three berries red
And pluck them from the stalk,
And burn them at the first cockcrow
That my spirit may not walk.

Elizabeth Siddal, *At Last*

Of all of the spring-blooming flowers, the cuckoo pint is perhaps one of the most unusual. Rather than bearing the showy, colourful flowers that we associate with the season, this plant is colloquially known in some areas as 'Lords and Ladies' due to its dual-sexed nature. It grows a large flower spike from its centre—the 'male' part, known as a spadix—protected by a pale green modified 'female' leaf known as a spathe. The 'pint' particle of the name is short for 'pintel', the old English for a penis. It grows best in wooded areas or along riverbanks, and is a good indicator of healthy and nutrient-rich land. For this reason, Germans believe that where it flourishes, the spirits of the wood are plentiful and content.

* Robert Chambers; *Popular Rhymes of Scotland*
† Daiva Šeškauskaitė; *The Plant in the Mythology*

The plant relies on flies to pollinate it, and so to attract them the tiny, clustered flowers that grow about the spadix spread the smell of rotting meat. The plant itself also heats the air around it to make it more appealing; it can increase the temperature surrounding it by up to 15°C, a trait that has been noted in studies of the plant as far back as 1777.

In the autumn, the spadix grows clusters of bright red berries. These berries, and the root of the plant, contain arione, an acrid toxin that gives the plant a bitter taste and causes long-term burning and blistering of the skin. However, this caustic nature hasn't stopped it from being used in the name of fashion for hundreds of years: the roots in particular are a reliable source of starch, and were lauded throughout the Elizabethan and Jacobean periods as being one of the best for stiffening lace ruffs and other linens. Though preferred by the nobility, it was likely hated by the laundresses who had to handle it, as its acidic nature would blister and burn their hands.

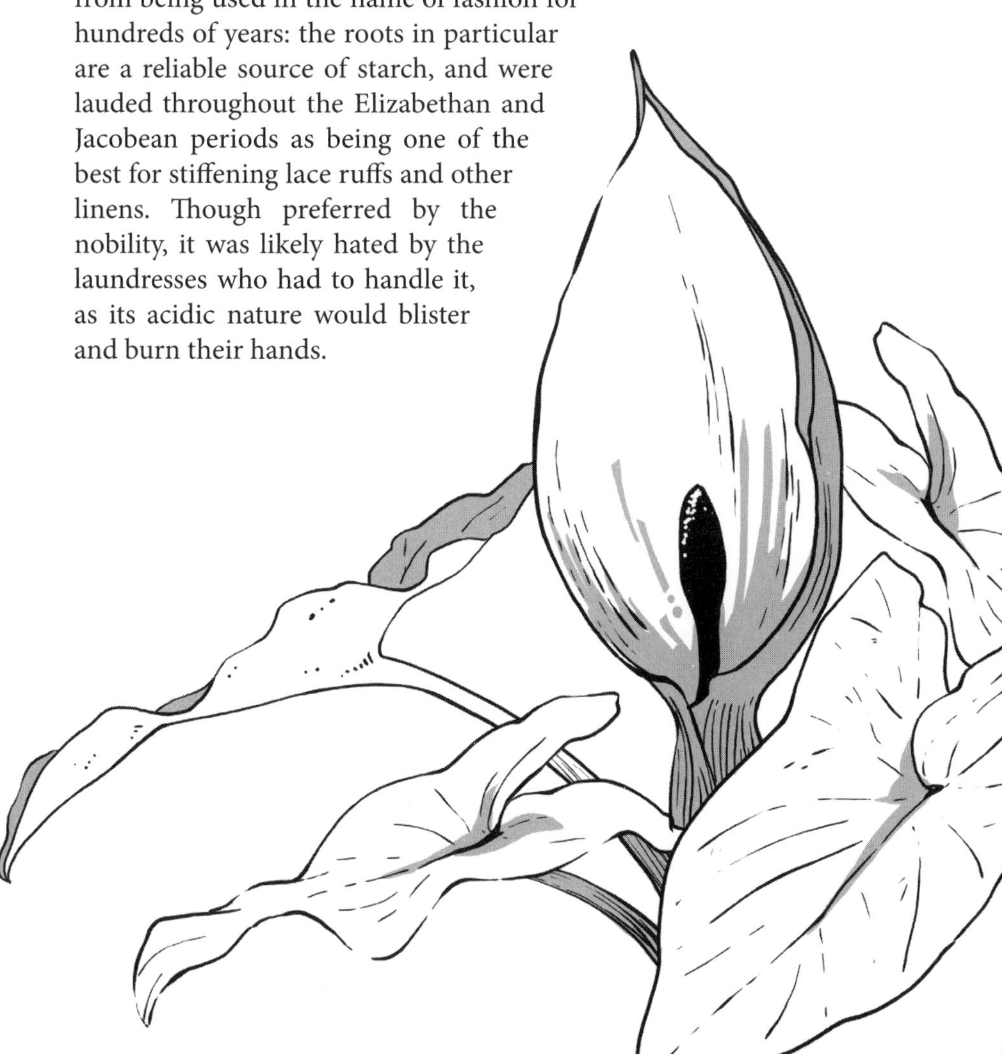

In earlier centuries, it was applied even more liberally in the search for beauty. The *Trotula*, a medieval manuscript that extolls various beauty treatments for European women, extols cuckoo pint as the ideal exfoliant to deal with coarse skin and to ultimately leave it whiter. Fortunately, it was advised to soak the root for five nights, disposing of the water each morning, to avoid causing lesions; this treatment by no means would have removed the plant's toxic nature entirely, but surely saved many women from burning themselves.

The search for beauty throughout history has brought women—and men—into contact with numerous plants and solutions that must have brought as much discomfort as they did beauty. As well as the famous tales of women who would drop belladonna extract into their eyes to dilate their pupils, sorrel juice was a popular treatment to soften hands and remove freckles, though the oxalic acid in the plant would have done more damage in the long-term. And in 1896, in *Youth's Education of Home and Society*, a soap was recommended to maintain healthy skin, containing, amongst the other ingredients, a quarter ounce of oil of bitter almonds… in other words, cyanide.

Another member of the Arum family, the dragon arum (*Dranunculus vulgaris*) is visually similar to the cuckoo pint, but has a deep purple spathe and an extended spadix that resembles a dragon's head. It was believed by both the Romans and the early Anglo Saxons that the root of this plant, when drunk with warm wine, would cure snakebites, and even in the 16th Century it was still used as a protection against snakes: 'Serpents wil not come neere unto him that beareth dragons about him.'*

* John Gerard; *Great Herball*

DAFFODIL: Narcissus pseudo-narcissus

When a daffodil I see,
Hanging down his head t'wards me,
Guess I may, what I must be:
First I shall decline my head;
Secondly I shall be dead;
Lastly, safely buried.

Robert Herrick, *Divination by a Daffodil*

These cheerful harbingers of spring are often one of the first flowers to bloom, frequently appearing even when there's still snow on the ground, and stay in flower for the majority of the season.

The name 'daffodil' is of uncertain origin. First recorded in use in 1592, it is thought to derive from the original Greek *asphodelus*, which is the name given to the asphodel, a cousin from the same lily family. It is thought that the 'd' may have been attached as the plant made its way to England through Europe, as the flowers may have been marketed as *d'asphodel* and later misremembered as 'daffodil'.

From an equally-disputed origin is the scientific name *Narcissus,* which is most commonly thought to come from the Greek myth about a man of the same name. Narcissus was a beautiful young actor who fell so much in love with his reflection in a pool that he wasted away and died, and was turned into the flower that we now know as a daffodil. However, it has been suggested that the name comes instead from the Greek *narcao,* meaning 'to become numb'; the same root word for the English *narcotic.* The association with *narcao* is no doubt due to the narcotic effects of the plant: just the scent of daffodils in an enclosed space can cause headaches and vomiting.

It's not just the scent of daffodils that is dangerous, but the bulb too. The bulbs contain a chemical called lycorine, which if ingested causes paralysis of the central nervous system, leading to collapse and eventual death. This dangerous association hasn't gone unremarked upon; in England in the late 1800s, it was believed to foretell of a bad future if the first daffodils of the year were seen with their heads hanging towards you.

The daffodil was also associated with death in the Greco-Roman world. It was said that daffodils were a favourite of Persephone, and that it was with these that Hades lured her into the Underworld. As a result, she continued to grow them on the banks of the Acheron. Graves were also within Hades' domain, and so daffodils were often grown on Greek burial mounds. It wasn't just the Greeks; daffodils were a common inclusion in early Egyptian funeral wreaths.

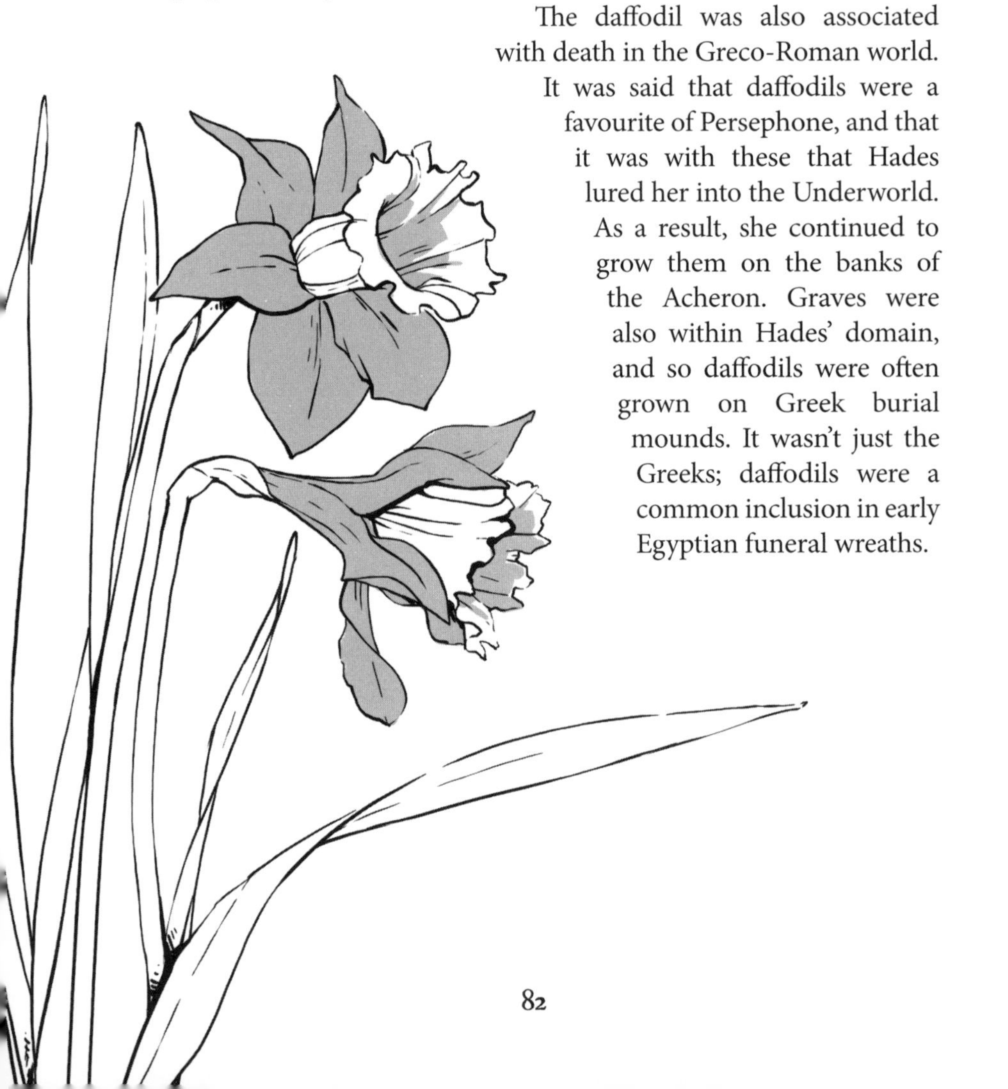

DARNEL: Lolium temulentum

> The fruitful soil that once such harvest bore,
> now mocks the farmer's care, and teems no more.
> And the rich grain which fills the furrow'd glade,
> rots in the seed, or shrivels in the blade;
> or too much sun burns up, or too much rain
> drowns, or black blights destroy the blasted plain;
> or greedy birds the new-sown seed devour;
> or darnel, thistles, and a crop impure
> of knotted grass along the acres stand
> and spread their thriving roots thro' all the land.
>
> Ovid, *The Metamorphoses*

Darnel is a ryegrass common across the world, and a great pest to farmers everywhere. It grows prodigiously in ploughed land where wheat is sown, and is so visually similar to the crop that in some areas it is referred to as 'false wheat'. Though the plant itself is not toxic, it's notoriously vulnerable to the ergot fungus, which if accidentally ground with wheat into flour can cause drunken effects such as giddiness and confused perception, trembling limbs, vertigo, and loss of strength, followed by blindness and severe illness. It can also cause the sensation of burning limbs, known as St Anthony's Fire. The latin name *tementulum* comes from this effect, *tementulus* meaning 'drunk'. The French and German names for this plant, *iyraie* and *Schwindel*, also share the same meaning.

Despite its dangers, the intoxicating effects of darnel poisoning were for a while taken advantage of in poorer areas of France and Germany as a way of getting drunk inexpensively, with beer being watered down and then fortified with darnel as a cheap narcotic intoxicant. Darnel wasn't the only plant used in this way; henbane and the levant nut are similarly-toxic plants that were added to beer before laws were passed against the practice in 1516.

A Norwegian legend tells of how darnel, known locally as *svimling* ('dizzy weed'), was used to sustain families through times of famine. During one such period in Kråkerøy, people were resorting to stripping bark from the trees to grind it into their flour to make it go further. The story tells of a woman whose children would not stop crying for food, and so she would make them darnel soup, so that they would fall into a stupor and sleep.* This story likely dates from the early 1800s, when massive crop failures coincided with the Napoleonic blockade, which stopped any English goods from entering Europe.

Though modern machinery now allows farmers to easily separate darnel grain from wheat, historically the fear of ergotism was a very real one. LSD, which wasn't processed as a drug until 1938, was originally derived from the ergot fungus, and the symptoms of ergot poisoning were both terrifying and often unavoidable.

Between the 15th and 16th Centuries, European history is rife with people displaying these symptoms: hallucinations, convulsions, paralysis, and dementia. Even the cattle that accidentally grazed on darnel would stop producing milk. This sickness was most commonly recorded in the damper river areas of south-western Germany and south-eastern France, where rye was a staple crop and conditions were perfect for ergot to thrive. But, without a greater understanding of ergotism, many of these people were thought to be possessed or under the influence of witchcraft. It is thought that the notorious 1692 witchcraft trials in Salem, Massachusetts, were the result of a town-wide infection of ergotism. Though this suggestion is still controversial in some circles, the damp season that preceded the incident was ideal for ergot growth, and the following summer—recorded to be much drier—ended both the ergotism blight and the bewitchments.

* Reimund Kvideland and Henning Sehmsdorf; *Scandinavian Folk Belief and Legend*

DEADLY NIGHTSHADE: Atropa belladonna

A rich pansy it was, with a small white lip
And a wonderful purple hood;
And your eye caught the sheen
Of its leaves, parrot-green,
Down the dim gothic aisles of the wood.
And its foliage rich on the moistureless sand
Made you long for its odorous breath;
But ah! 'twas to take
To your bosom a snake,
For its pestilent fragrance was death.

John Boyle O'Reilly, *The Poison Flower*

Of all the poisonous plants in the world, the one with the most sinister reputation must surely be the deadly nightshade. Also known as belladonna, this member of the Solanaceae family is closely related to mandrakes, tomatoes, chilli peppers, and potatoes. It is also one of the oldest recorded botanical poisons, referenced even so far back as the Ebers papyrus, written in 1550BC.

The plant grows in damp, shady spots, such as woodlands and riversides. The Vale of Furness in Lancashire, England, is particularly renowned for its presence; it is known locally as the Valley of Nightshade, and sprigs of nightshade are a common motif carved on the seals of Furness Abbey's ruins. It also grows freely in Romania, where it is so respected and beloved that it has been given a wealth of titles: *Cinstita* ('The Honest'), *cireașa lupului* ('wolf's cherry'), *Doamna Codrului* ('Lady of the Forest'), and *Împărăteasa Buruienilor* ('Empress of Weeds').*

The berries, even when green, are remarkably shiny, and noted by herbalist John Gerard as 'a berry of a bright shining black colour, and of such beauty as to allure any to eat thereof'. Despite their beauty, eating the berries is not advised; though eating one might not bring death, one berry on a plant can be up to 50 times more toxic than the

* Alexandru Borza; *Ethnobotanical Dictionary*

one right next to it, and what may not kill the first time certainly may the second. Unlike most toxic fruits they are deeply and seductively sweet, and for this reason are seen as evil, as any dangerous plant with good manners was supposed to grow its berries bitter to deter ingestion.

Nightshade is high in atropine and scopolamine, which are both so toxic that even the smallest amount can cause psychosis, hallucinations, convulsions, and seizures. Even simply touching the plant can raise blisters on the skin. A common rhyme used to describe the effects of atropine poisoning is: 'Hot as a hare, blind as a bat, dry as a bone, red as a beet, and mad as a hatter'. An association with madness isn't uncommon: in 1555, botanist Andrés Laguna wrote of nightshade that 'a dram of extract from the root, when dissolved in wine, produces fleeting images that please the senses. But if the dose be doubled, it drives a man mad for three days.'

Deadly nightshade is named *Atropa* in honour of the third Greek Fate, Atropos. Atropos (The Unturnable) was the eldest of the three Moirai, the Fates, the arbiters of the life and death of mankind. Before her came Clotho (The Spinner), who spun the thread of a mortal's lifespan; then Lachesis (The Allotter), who measured the thread and the length of that person's life; and Atropos, when the time came, would cut the thread with her shears. It was said that Atropos took the form of a nightshade when she was in the realm of the living.

The nightshade's

deadly association is well earned; it has been responsible for countless deaths throughout history, whether intentional or not. It is rumoured that the Roman Emperor Augustus was killed by his wife, Lucia Drusilla, by way of a plate of nightshade-laced figs, and more recently was the cause of death of noted witch revivalist Robert Cochrane in 1966.

Well-known are the tales of medieval Venetian ladies using drops of atropine to dilate their pupils to appear more beautiful. However, using it too frequently could allow it to travel along the optic nerve and lead to madness. Walking around with dilated pupils would also make it difficult and painful to see in the daytime, though at night, it might just have improved their eyesight! But the ability to enlarge the pupils wasn't just fashionable at the time; it was also made use of by early opticians prior to surgery to make their work easier, and was still used by opticians until only a few decades ago.

Popular belief tells that the name *belladonna*—beautiful lady—comes from this practice of enlarging the pupils for beauty purposes. However, there is no proof that this fashion ever travelled beyond Venice, and it has been proposed that the name may alternately come from *buona donna,* or 'good girl', the name given to the Italian witch doctors that the poor relied on when they could not afford access to a physician.* It was believed that

* A Brighetti; *From Belladonna to Atropine, Historical Medical Notes*

these witches became such by inheriting the power from another. Once inherited, the witch could not die until she found someone else to transfer it to. In *Gypsy Sorcery and Fortune Telling,* Charles Leland retells a legend he was told in 1886 in Florence:

There was a girl here in the city who became a witch against her will. She was ill in a hospital, and by her in a bed was an old woman seriously ill, yet who could not die. And the old woman groaned and cried continually, 'Alas! to whom shall I leave?'—but she did not say what.

Then the poor girl, thinking of course she meant property, said: 'Leave it to me—I am so poor.' At once the old woman died, and the poor girl found she had inherited witchcraft.

Atropine was also used more recently in 2018, when the nerve agent Novichok was used to poison two Russian expats in Salisbury, UK. Effects of Novichok include muscular spasms that disrupt the heart and cause respiratory arrest. In the case of the 2018 attack, atropine was used to return the victims' hearts to a normal rate.

With such a reputation for death, it was only so long before nightshade became associated with magic and otherworldly mischief, and the superstitions that surround this plant are numerous. Henry G Walters, a professor and plant researcher in the early 1900s, believed that all plants were capable of loving and creating memories, and that they might hold a grudge in the way that lovers do. He believed that the deadly nightshade was a plant full of hatred. In Normandy, it was said that any who walked barefoot over it would suffer immediate madness,* and in the Highlands of Scotland, it was said to grant the ability to see ghosts.† The Irish thought it more easily used for evil; the distilled juice, if drunk, would make the drinker susceptible to whatever they were told.‡ This latter may have some element of truth to it: scopolamine, one of the chemical compounds found in nightshade, is used as an ingredient of some truth serums. It is a hypnotic that

* William Branch Johnson; *Folk tales of Normandy*

† James Kennedy; *Folklore and Reminiscences of Strathspey and Grandtully*

‡ Jane Wilde; *Ancient Legends, Mystic Charms, and Superstitions of Ireland*

alters higher cognitive function, and has been used in the USA since 1922, where it has played a part in the conclusion of multiple court cases. Despite serious questions having been raised as to the reliability of these tests, they are still in use today.

But most notably, nightshade became known in tales as a plaything of witches and the Devil. It was said to be beloved of and tended by the Devil himself, who left it alone only on Walpurgis Night (April 30th to May 1st) when he retired to prepare for the witches' sabbat. On this night, you would be safe to dig up the roots of the plant; but the Devil would leave behind a 'nightmare monster' to protect it, which could be appeased only by offerings of fresh bread.

Belladonna was one of the plants—alongside opium poppy and monkshood—believed to be used by witches in their 'flying ointment' that would transport them to the Black Sabbat. The ointment was not actually used for flying, but was simply a name for a mixture used to encourage hallucinatory dreaming—a side effect of ingesting scopolamine and opiates.

Though not as world-famous as the deadly nightshade, a close relative, the black nightshade (*Solanum nigrum*) is responsible for just as many historical misadventures. Unlike its cousin, its toxicity can be somewhat unreliable. In 1794 was recorded a case where an entire family misidentified and ate it, and though the mother and child fell ill the father did not. Many members of the Solanum genus can be difficult to predict in this manner; the fruits are edible when ripe, but toxic when unripe, and all other parts of the plant are always toxic.

It features in Shakespeare's *Macbeth,* when the Scottish king Duncan is under assault by Sweno, the king of Norway. Macbeth invites Sweno and his men to a dinner to discuss their terms of surrender, but laces the meal and drinks with black nightshade. Once they are asleep, Macbeth brings his men in to slaughter the enemy soldiers, and Sweno only escapes with the help of those who abstained from drinking.

DEVIL'S BIT SCABIOUS: Succisa pratensis

Endless lanes sunken in the clay,
Bays, and traverses, fringed with wasted herbage,
Seed-pods of blue scabious, and some lingering blooms;
And the sky, seen as from a well,
Brilliant with frosty stars.
We stumble, cursing, on the slippery duck-boards.
Goaded like the damned by some invisible wrath,
A will stronger than weariness, stronger than animal fear,
Implacable and monotonous.

Frederick Manning, *The Trenches*

Devil's bit is a tall herb native to Europe, though it now grows across North America and central Asia. Thriving in grassland and heath areas, it is a good source of nectar to pollinators and a core food source of rare insects such as the marsh fritillary butterfly and the bee hawk moth.

The name *scabious* comes from its historical use in treating scabies and other skin afflictions, most notably the sores that were caused in times of bubonic plague. This healing property came from the root, which is short and blackened, and believed to have a great many medicinal values. Folk legend tells that the Devil, out of spite to mankind, bit the root short to stop it from being so useful. The plant bears a similar name in Germany; the *Hortus Sanitatis,* an early German herbal published in 1491, calls it *Morsus Diaboli,* the Devil's bite.

In the south of England, scabious is seen as a mischievous

plant with the ability to cause unexpected fires. It seeds well in ploughed fields and continues to grow long after most other plants have been harvested and dried, which means that the moisture in the leaves can begin to ferment and then combust, setting fire to hay barns and bales left out in fields. To make sure that it is safe to store hay, farmers will take the plants—colloquially known as fire leaves—and twist them tightly. If water is squeezed out, it is not yet safe to store. Another plant that poses similar dangers is the hoary plantain, which grows alongside devil's bit.

A related plant, the sweet scabious (*Scabiosa atropurpurea*) is also known as the Mournful Widow. In Portugal and Brazil, it is commonly included in funereal wreaths, and is locally called *saudade*: a word that has no direct translation in English, but refers to a state of profound melancholic longing for something or someone that is beloved but missing. It is usually used with the unspoken knowledge that the thing or person who is missed might never return.

DOGBANE: Apocynum spp.

The lesser Periwinkle's bloom,
Like carpet of Damascus' loom,
Pranks with bright blue the tissue wove
Of verdant foliage : and above
With milk-white flowers, whence soon shall swell
Red fruitage, to the taste and smell
Pleasant alike, the Strawberry weaves
Its coronets of three-fold leaves
In mazes through the sloping wood.

Bishop Richard Mant, *Poems*

Dogbanes are a small family of flowering plants that grow widely across the globe, mostly in tropical or subtropical areas. Members of this family grow in many varied ways—they can appear as trees, as herbs, or as vines—but they have one similarity: they all have a poisonous milky sap which can cause swelling and inflammation. This only seems to affect certain people, though, and some can come into contact with it with no repercussions.

The plant is named dogbane for its historical use in killing not just dogs but also other predatory pests such as wolves and foxes. 'Bane' comes from the old English *bana* and previously the Norse *bani,* both meaning 'slayer' or 'murderer'.

Common to Canada and North America, one particular member of this family is called the flytrap dogbane (*Apocynum androsaemifolium*) for the way that it ensnares unwilling pollinators. The small, bell-shaped flowers are filled with a sweet nectar, but the anthers are specifically shaped so that a hungry fly or ant must squeeze past them to feed. The insides of the plant are so complex that the prey is then entangled within it, and by the end of summer the flowers are filled with dead bodies.

One of the few non-poisonous members of the dogbane family is the periwinkle (*Vinca major* and *V. minor*). Despite its lack of toxicity, in Europe it has become associated with ghosts, witches, and the dead, and carries such folk names as sorcerer's violet (France), hundred-eyes (Italy), and flower of immortality (Germany).

In Wales it is simply 'the plant of the dead'. Grown chiefly on graves, it is seen as unlucky to uproot it, as the spirit of the person within the grave would then haunt the dreams of the one who had done the deed.* In other countries, it is seen as a guardian of the dead, to be respected rather than feared, and is woven into wreaths to be placed on coffins, mostly those of children.

In medieval England, a criminal on their way to the gallows would be marked with a garland of periwinkles about their neck.† The reason for this choice of flower is unknown, but perhaps the connection with the noose and the grave is enough to have started the trend.

DOG'S MERCURY: Mercurialis perennis

They be tiddy critturs, no more than a span high, wi' arms an' legs as thin as thread, but great big feet an' hands, an' heads rollin' aboot on their shoulders.

E. Rudkin, *Folklore of Lincolnshire*

Dog's mercury is a fast-spreading weed that is pervasive across much of Europe and the Middle East. An indicator species of ancient woodlands, it prefers shady, damp areas and is made distinctive by

* Marie Trevelyan; *Folk-Lore and Folk Stories of Wales*
† William Emboden; *Bizarre Plants: Magical, Monstrous, Mythical*

its toothed, spear-shaped leaves, small green clusters of flowers, and rotten smell. Unlike the dogsbane family, the 'dog' in this plant's name does not come from any correlation with animals, but instead meaning 'false' or 'bad'. A 'dog' plant is one that has no medicinal uses at all, and in this case may have been named to separate it from annual mercury, which is visually similar and does have medicinal value.

A member of the spurge family, which is notorious for producing poisonous plants, all parts of the dog's mercury are highly toxic. It contains trimethylamine, which is also present in hawthorn blossoms, and is the cause of the plant's rotting flesh smell. It also contains the poisonous mercurialine, which can cause internal inflammation, muscular spasms, nausea, and drowsiness. Though death from encountering this plant is rare, an account from 1693 tells of a family of five who boiled and ate dog's mercury and fell ill, with one of the children dying as a result of the incident.*

Another folk name for dog's mercury is the boggart's posy. The boggart is a malevolent and mischievous *genius loci*—the protective spirit of a specific place, such as a home, a river, hill, or other geographical feature. Also known as a bugbear or bogeyman, the name comes from the Irish *púca*, 'ghost', a mythical creature similar to the boggart that still exists in Ireland today. In Scotland, it is known as a bogle.

Boggarts would choose a particular household to haunt, and would follow its family even if they moved away. Unlike wild boggarts, which would abduct children and drown strangers, domestic boggarts would turn milk sour, put clammy hands on a sleeper's face at night, and steal their bed sheets. And this was a boggart on its best behaviour—

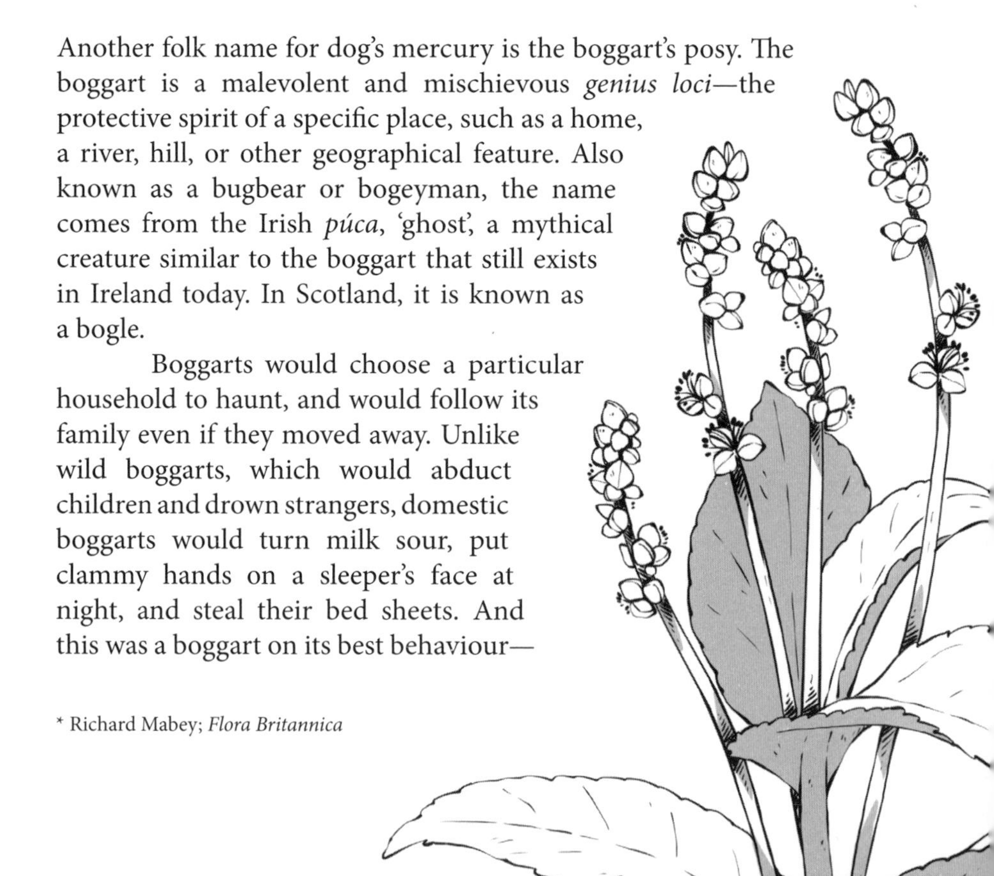

* Richard Mabey; *Flora Britannica*

it was only when you gave it a name that it would become truly unruly and destructive. The only method for doing away with its mischief was to hang a horseshoe on the door, or to leave a pile of salt outside the bedroom. It is thought that domestic boggarts might once have been silkies, a species of helpful household sprite, which had been offended or otherwise ill-treated and turned vicious. Boggarts are commonly said to be humanoid in shape, though usually bestial and ugly; one was said to be a squat, hairy man with unusually long arms, and two other reports, from Yorkshire and Lancashire, claimed the boggarts in question appeared as horses or made the sound of baying hounds as they chased their victims.

DRAGON'S BLOOD TREE: Dracaena cinnabari

"Death on the gods of death!
Over the thrones of doom and blood
Goeth God that is a craftsman good,
And gold and iron, earth and wood,
Loveth and laboureth.

G. K. Chesterton, *The Ballad of the White Horse*

The dragon's blood tree is a plant endemic to Socotra on Yemen, where over 30% of the island's plant species occur nowhere else on earth. The entire tree looks much like a mushroom, with the trunk and undersides of the branches completely bare, and only the ends of the branches tipped with leaves and flowers. Once a year, tiny orange fruits cluster between the branches.

What is most remarkable about the dragon's blood tree, however, is also the reason for its name. The sap of this tree is bright red, and freely 'bleeds' when the living wood is cut into. This resin dries into a solid crystalline form that, throughout history, has been used in dye, medicine, furniture varnish, alchemy, and numerous other applications. Legend supposes that the first tree grew from the blood of a dragon that had perished in mortal combat with an elephant. This is what also gives the tree its scientific name, *Dracaena,* coming from the Greek *drakaina,* a female dragon.

There are a number of 'bleeding' trees across the world, and though most modern sources of dragon's blood come from the *Daemonorops draco* (the rattan palm) in South East Asia, the original source of this product was the *D. cinnabari.* Though the resin from multiple bleeding trees can and has been marketed as dragon's blood, we know that *D. cinnabari* was used by the early Romans as a dye, painting pigment, and early medicine for respiratory diseases. Both *D. cinabarri* and *D. draco* (the draconis palm) were used as a source of varnish for furniture and later 18th Century violins. In the same period, there was a recipe for toothpaste that required dragon's blood. In current-day American Hoodoo, dragon's blood is still used as an incense, and also to make a dye known as 'dragon's blood ink', which is used for inscribing talismans and magical seals.

Another of the trees that produces a 'bleeding' sap is the *Croton lechleri*, the Sangre de Drago. The resin of this tree was used as a dark red dye for Aztec textiles, and there still exists a legend about how it came to exist:

The legend tells of a prince who wore only the finest gold and gems. Hungry for more wealth, he hired a gang of thieves to waylay wealthy merchants and steal their wares. He would then take a slave into the forest to dig a hole so that he might bury the stolen jewels beneath a tree; but once the hole was dug, he would kill the slave and bury their body with them. Not only would the location remain secret, but the slave's ghost, so he planned, would protect his hoard forever.

This continued for several years, and rumours of his deeds spread amongst his slaves. Finally, judgement caught up with him: on one such trip out into the jungle his slave turned on him before they could be slain, and murdered the prince instead. The prince was finally the one to be buried, and the slave found a better life with the gold from the hoard.

*From the place where the prince was buried began to grow a tree that wept blood, and this is how blood trees are said to exist in the world.**

In La Orotava, a town in Tenerife in the Canary Islands, there was a tree that was known as the Dragon-Tree of Orotava, and was worshipped as sacred by the Guanches, the original inhabitants of the island. Before the great redwoods of California were known of, this tree (a *D. draco*) was considered to be the greatest and tallest of living trees, at eighty-two feet tall and with a girth of seventy-five feet.† Approximately 6,000 years old, this tree had been hollowed out to create a small sanctuary for religious purposes. Unfortunately, the tree was felled by winds in 1867.

After this, the next specimen to take the title is known as *El Drago Milenario* (The Thousand-Year-Old Dragon) and grows in Icod de los Vinos, also in Tenerife. It is sixty-five feet tall and is thought to be between 800 and 1,000 years old.

Finally, another bleeding tree of note is the *Pterocarpus* genus. Native to the Sahelian region of West Africa, these trees are also called the

* Sheryl Humphrey; *The Haunted Garden: Death and Transfiguration in the Folklore of Plants*
† Alexander von Humboldt; *Cosmos: A Sketch of a Physical Description of the Universe*

African teak, or the kino tree. This latter name comes from the same bleeding effect that gives the other dragon trees their names; but this time it is not a resin that bleeds, but a botanical gum called kino, which is thinner and more liquid than resin. This kino is still used in tanning and dyes, and as an aphrodisiac.

An interesting account of the superstition that surrounded one local tree was recorded in *The Society of Malawi Journal*, 1978. The account was taken from local villager Redson Ng'ambi in 1966, and is as follows:

'There is a tree at Namwafi Village beside the road. If you cut one branch off, you will die. It is a very big tree with many branches. If you cut it, red blood begins to flow out and if you climb it, you will never come back down.

There are some wizards inside it. If you pass by it you can hear them. Even if you are very far away you can hear the noise. If you listen to the noise you will find that many people are surrounding you. Some of these people will beat you, and at night you will die.

One day it was raining, and thunder and lightning burned it. It moved about eight feet. All the leaves in Bulambia had fallen off the tree, and there was a heavy wind which killed many living things, such as birds and chickens.'

With stories like these, it's not surprising that such a tree might be seen to be haunted or magical in some way. However, it may still be possible to untangle the truth behind this particular tale. The tree in question grew near Namwafi (properly referred to as Mwenebwib'a) on the Kaseye River, not far from Chinunka in Malawi. Being near a river, the wind blowing through a river valley no doubt could make the sound of voices or moaning—the aforementioned wizards—and if such a storm really had ravaged the area, perhaps the tree itself did not move, but instead the river beside it may have been rerouted.

DUMB CANE: Dieffenbachia spp.

There is a muteness - the tocsin bell
Has made us close our lips.
In our hearts, once so ardent,
There is a fateful emptiness.

Aleksandr Blok, *Those Born in Obscure Times*

Originally from South America, the dumb cane—also known as the leopard lily or mother-in-law's tongue—has found its way into many homes as a popular ornamental houseplant. Though commonly owned, less well-known is the origin of the name dumb cane; it originates from the plant's poisoning effect that inflames the vocal cords, causing muteness and an inability to breathe.
This is caused by microscopic needles of calcium oxalate, called raphides, which are present in all parts of the plant, but particularly the stalks. These needles are irritating to the skin, and when imbedded in the soft tissues of the mouth and throat cause intense burning, salivation, and swelling of the lips, tongue, and mouth.

Though it has never been recorded as responsible for the death of a person, it was once used on Caribbean sugar plantations as a punishment for slaves who had disobeyed orders. They would be forced to eat the leaves of the plant, which would cause them to be rendered mute and unable to eat for some time.*

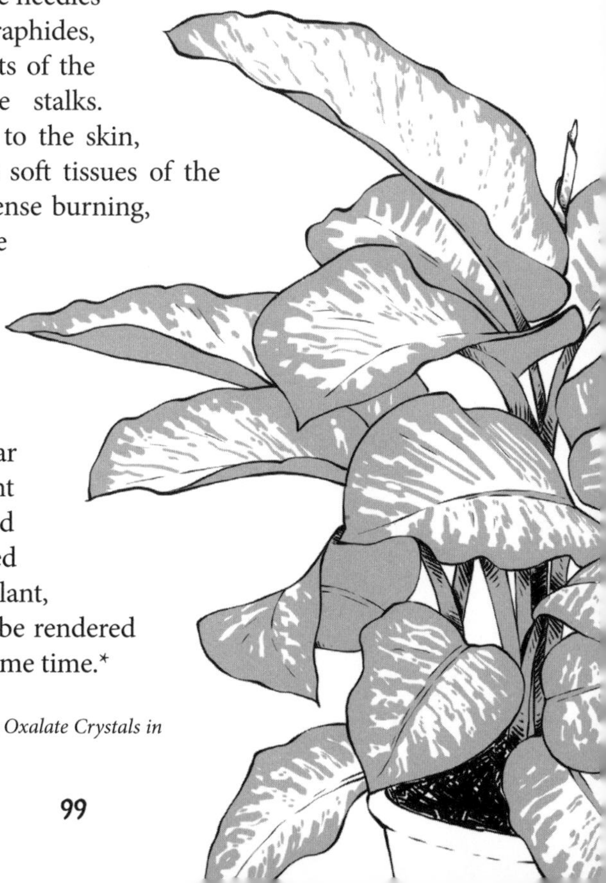

* Hui Cao; *The Distribution of Calcium Oxalate Crystals in Genus Dieffenbachia*

ELDER: Sambucus spp.

Elder that hath tough bark, tree that in truth hurts sore;
Him that furnishes horses to the armies from the sidhe burn
so that he be charred.

Unknown author, *The Violent Death of Fergus mac Léti*

Widespread in the Northern Hemisphere, the elder tree is a common sight growing wild in hedgerows, and a welcome one in early spring for its cheerful masses of white blossoms. Elderflower and elderberries are increasingly sought out as a culinary flavourings, and the wood, which is easily hollowed, has been used in crafting for hundreds of years. It also produces multiple dyes; blue and purple from the berries, yellow and green from the leaves, and black from the bark.

But despite its uses, elder is, like the blackthorn, a tree that has somehow become burdened with superstition and bad luck. It's likely that much folklore regarding these two trees has become crossed over due to their similarities; they are both common hedgerow trees of about the same size, both spring-flowering, and both end the year laden with dark, edible berries. A good example of this potential crossover is in a 1905 issue of *Folk-Lore*, whichdiscusses a gamekeeper who tripped on an elder, and died of tetanus from the thorns. The article specifically mentions a superstition that a wound from elder is fatal, but the elder tree does not bear thorns in any manner, and the deadly superstition most commonly belongs to the blackthorn.

Whether confused with blackthorn or not, the fact remains that the elder is, at the end of the day, a cursed tree. In the British Isles, falling asleep beneath an elder assured that you would never wake up again; it is also associated with witchcraft, and as a result it was said to be unlucky to cut or touch the wood after dark.* This belief is also held by the nomadic Romani, perhaps due to the fact that elder wood burns fast and hot, with a great deal of noise and a fearsome smell. The smoke of a burning elder branch is said to reveal witches, and a child both baptised and anointed about the eyes with the juice of the bark was said to be able to see and converse with them.

The dwarf elder (*Sambucus ebulus)* was said to grow only where blood had been shed, particularly the blood of the Danes. This superstition is shared with the Pasque flower (*Pulsatilla vulgaris)*, as both grow well on open and elevated land, such as the raised boundary banks of old hill forts where the Danes fought. The toxicity of the unripe berries was even blamed on an old curse placed by the Danes who had fallen in battle.† The Welsh call it *Llysan gwaed gwyr*, or 'plant of the blood of man', and in England it is called bloodwort or deathwort, 'wort' simply coming from the old English *wyrt*, or plant.

Not far away, in Ireland, there is a saying: *Trí comartha láthraig mallachtan: tromm, tradna, nenaid.* 'There are three tokens of a cursed site: the elder, a corncrake, and nettles.' A cursed site was believed

* Enid Porter; *Cambridgeshire Customs and Folklore*
† Berta Lawrence; *Somerset Legends*

to be any area barren of life, and perhaps there is some truth to the observation. The elder is one of the fastest trees to take advantage of recently-cleared ground, as do nettles, and the lonesome corncrake prefers shallow scrubland such as hayfields and grassland.

Another Irish belief was that the elder had in it a bad temper, or a mischief. It was said that if a person were to be slain with a weapon made of elder wood, after their death their hand would appear rising from their grave. This is just one of many superstitions that connects it to the dead: the Scottish planted it on graves to prevent the dead from rising, a custom mirrored in historical Tyrol (now part of northern Italy and western Austria), although the Tyroleans believed that the sight of elder blooming on a grave signified that the dead person had travelled to paradise.

Elder has also been found mentioned in instructions on how to create a *Corp Chreadh*—a clay corpse, or poppet. Following is an excerpt from a 1566 English record:

As for pictures of claye, their confection is after this manner: ye must take the erthe of a newe made grave, the ribbe bone of a dedde man or woman ... and a black athercobbe (spider) *with also the inner pithe of an elder* (tree) *tempered in warm water wherein todes* (toads) *must be first washed.**

This clay poppet would then be stuck with pins, or otherwise placed in a running stream, where the water would begin to crumble and waste away at the *corp*, and therefore the intended victim as well.†

Across Europe is a web of connections between gods and the spirits that favour this tree.

The Norse *Poetic Edda* talks of *Svartálfar,* 'black elves' who lived on the coast and in caves who chose the elder tree as a place for their rituals, as they particularly liked the strong scent of the flowers. There is also mention of the Elder Mother, a dryad who dwelled within and watched over the tree. Anyone who harmed the tree would

* Marion Gibson; *Witchcraft and Society in England and America,* 1550-1750
† Robert Craig Maclagan; *Notes on Folklore Objects Collected in Argyllshire*

be punished by her; a nod back to similar European beliefs that the tree was possessed and guarded by witches. This belief passed over to England, where she is known as Lady Ellhorn, and Germany and Denmark, where she is named *Hylde-moer* or *Hylde-yinde* (Elder-Mother or Elder-Queen). In Lower Saxony (Germany), woodmen would be expected to go so far as to ask permission before cutting wood from the elder, and would bend a knee and ask: 'Lady Elder, give me some of thy wood, then I will give thee, also, some of mine when it grows in the forest.'*

In Germany, the holes in the tree are said to be also favoured as doorways by the *Waldgeister* (literally 'forest ghost'), an ancient race of wood sprites who protect the tree. Though there are stories of them leading lost travellers back to the road, they are also thought to enact the Elder Queen's revenge on those who damage the tree, often making them fall asleep or disappear entirely.

Moving further east to the coast of the Baltic Sea, the gods are even further involved. When the historic state of Prussia existed, the elder was the home of Pušaitis, the god of the earth. Twice a year, farmers would leave bread and bear meat at the base of the elder tree, so that his servants, *barstukai* (sprites that were closely related to the *kaukas,* the harvest hobgoblins), would help them with the harvests.† Both Pušaitis and his helpers were deeply respected, and seen as guardians of the forests.

In modern-day Lithuania, once part of the same area, the elder tree belongs to Velnias, or Velas, the god of the kingdom of the dead. He is also god of magic, metamorphosis, the creator of black animals and birds, and a guardian of forests. This latter connection may have come from the Lithuanian belief that the spirits of those who pass away live for some short time in the trees before moving on to the netherworld, and the sound of the leaves in the wind is the sound of the dead passing messages back to their loved ones. To cut down a tree is forbidden, and places where trees or tree stumps stand are seen as ill-omened places to build a house.

* Richard Folkard; *Plant Lore, Legends, and Lyrics*
† Daiva Šeškauskaitė; *The Plant in the Mythology*

FOOL'S PARSLEY: Aethusa cynapium

What god appointed for thee, little one, the burden of so dire a fate?
Scarce on thy life's earliest threshold, art thou slain by such a foe?
Was it that thus thou mightest be sacred for ever to the peoples of Greece
and dying merit so glorious a burial? Thou diest, O babe, struck by the
end of the unwitting serpent's tail, and straightway the sleep left thy limbs
and thine eyes opened but to death alone.

Publius Papinius Statius, *Thebaid, Book* V (translation by J. H. Mozley)

Fool's Parsley is an annual herb made distinctive by its tall, umbelliferous white flowers. It is related, and also visually very similar, to other poisonous plants such as hemlock, ground elder, giant hogweed, and water dropwort. These are all poisonous plants that thrive across the British Isles and the Eurasian landmass.

As a common-growing wildflower, it has a number of localised folk names, such as devil's wand, lesser hemlock, and mother-die. Regarding the latter name, there are at least twelve recorded European plants with the 'mother-die' moniker; it was generally said that it was taboo for a child to pick one of these, as it would surely bring about the death of their maternal parent. Despite the number of plants bearing the title, fool's parsley is the only one of these that is actually poisonous. However, cow's parsley is also named this, which suggests that maybe hemlock, the appearance of which is easily confused with cow's parsley, was the original intended recipient.

All parts of fool's parsley are toxic, and cause a burning heat in the mouth and throat. This can continue down into the stomach, causing blistering and burns, as well as muscle paralysis and death by suffocation. Though records of anyone dying from eating fool's parsley are rare, it's not unheard of: there are records of an English child in 1845 who mistook the roots for turnips and died of asphyxiation and lockjaw.*

The plant is known as 'fool's' parsley because of its striking similarity to the herb that we know more intimately, the true parsley. Unfortunately, the fool's toxic nature has given the common kitchen parsley something of a bad reputation, and for a long time now it has been associated with the Devil. Since parsley takes so long to germinate when grown from seed, it was said to grow all the way down to the Devil and then back up again. Uprooting the plant would open a path to hell, and a close family member would be in danger of finding themselves taken there.

It was also a funerary herb, and dedicated to Persephone, the queen of the Greek Underworld. This connection comes from the legend of Opheltes, the infant son of King Lycurgus. Left unattended by his nurse one day, he was killed by the bite of a serpent (or strangled by a serpent-like dragon, depending on the version of the tale). Where his blood ran to the ground, parsley sprang forth, and before he was

* Charles Johnson; *British Poisonous Plants*

buried Opheltes was renamed Archemorus, meaning 'the forerunner of death', as his early demise had been predicted by the Oracle at Delphi. Parsley continued to be associated with funerals from this point. Greek tombs would be decorated with wreaths of the herb, and the phrase *de'eis thai selinon,* 'to need only parsley', was used to suggest that someone was close to death.

FOXGLOVE: Digitalis purpurea

The Foxglove on fair Flora's hand is worn
Lest while she gather flowers, she meet a thorn.

Abraham Cowley, *Book of Flowers*

Common across the Northern Hemisphere, foxgloves are particularly distinctive in summer with their tall, bell-shaped flowers that can grow up to five feet in height, raising them well above most other wildflowers. Though wild foxgloves are typically purple in colour, varieties have been cultivated in pink, white, red, and cream. A black dye can be made from the leaves, and in Wales, the colourant is used to paint lines and crosses on stone cottage floors to keep witches away.

The one thing more abundant than the flowers on its stalks are the number of folk names—and legends—that it has gathered. In keeping with its English name, many of them relate back to foxes or gloves; but more often they revolve around fairies, as many plants with bell-shaped flowers do. One Norwegian legend tells how a fairy taught foxes how to ring the bells of the foxglove to warn each other of nearby hunters. In Norwegian the foxglove is known as *rev-bielde*, 'fox-bell'. There is a strong similarity to the Finnish tale of the bluebell, known in that country as 'cat's bell', where the bell-flowers would warn mice of approaching cats.

Another Norwegian folk name is *Reveleika,* 'fox music', in reference to an early instrument called a tintinnabulum, which consists of a ring of bells hung from an ornate arch. The similarity in appearance to the foxglove's tall flower stalks likely made this connection, and even

in early Anglo-Saxon England it was called *foxes-gliw,* with the same translation as the Norwegian.

Many of its folk names refer to gloves or fingers. In French it is known as *Gantes de Notre Dame* (Our Lady's Gloves) and *Doigts de la Vierge* (Fingers of the Virgin Mary), and in Wales it is *Menyg Ellyllon,* or 'goblin's gloves'. An early English name is 'folk's glove', perhaps referring again to the fair folk of the British Isles, as it was thought that the spots on the flowers came from where fairies would alight on the petals after flight. Another popular British tale tells how fairies gave the flowers to foxes to put on their paws, so that they would not make a sound when raiding a chicken house; hence, 'fox's glove'.

One outlier in colloquial names is on Guernsey, where the plants are casually known as a claque. This comes from the child's game of bursting (claque) the flowers by capturing air within the bells and then pressing the ends until they pop.

Despite the many legends, the truth of the name is a little less magical. The name was first recorded in the 1500s by botanist Leonard Fuchs, *fuchs* being the German for 'fox'. Since the scientific name, *digitalis,* comes from the Latin for 'finger' and refers to the way that the size of a flower is approximately the length of a finger, the connection to gloves was not far behind and the name became recorded as *foxes glofe,* 'fox's glove'.

In North Devon and Cornwall in England, the foxglove is associated with Saint Nectan. Born in 468AD, Nectan was the eldest son of the Welsh King Brychan, who left his family to become a hermit after being inspired by stories of Saint Anthony. Though he lived a peaceful life, one summer two passing robbers came across the pair of cows that Nectan owned, and stole them away. Nectan followed them, and when he caught up with the thieves they cut his head from his shoulders. Determined not to die away from his home, he simply picked up his head and walked back to his hut, where he finally lay down and died. Where his blood fell during that walk, there sprang forth foxgloves.

As previously mentioned, foxgloves have become closely associated with fairies. The fair folk were said to live within the bells, and when fairies would pass the flowers, they would bow their heads in respect; another trait shared with the similarly-shaped bluebell. It is particularly loved by the Shefro, a gregarious Irish fairy who wears the bells of the foxglove during its midnight revels.*

The foxglove also plays a part in ridding a family of a suspected changeling. A common myth in Europe, changelings are said to be fairy children that have been switched out with real human babies, or in some rare cases (such as the famous incident of Bridget Cleary in Ireland in the 1800s), an adult. The fear of changelings was very real, and has led to many cases of the murder and infanticide of supposed fairy usurpers, and likely came from early medieval concerns over children born with developmental disabilities and diseases. Whatever the reason for a child being suspected of being a changeling, there were a number of suggestions for how to force the fairies to return the original babe. One supposed method involved bathing the child in the juice of the foxglove plant, or to leave a piece of the plant beneath the child's bed.† A more elaborate 'cure' is as follows:

Take Lusmore [an Irish name for foxglove] *and squeeze the juice out. Give the child three drops on the tongue, and three in each ear. Then place it* [the suspected changeling] *at the door of the house on a shovel and swing it out of the door on the shovel three times, saying: 'If you're a fairy, away with you'. If it is a fairy child, it will die that night; but if not it will surely begin to mend.*‡

If this remedy was followed, the chances are that the child would have died more likely than it would have recovered. This is because foxgloves contain digitoxin, a poison that impedes blood circulation and slows the heart until it stops altogether. A common cause of death was by accidental poisoning; in *Time's Telescope* in 1822 an article discussed a rising fashion in Derbyshire in England of poor women drinking foxglove tea as a cheap means of intoxication, as 'it produces a great exhilaration of spirits, and has some singular effects on the system'. The

* Walter Evans-Wentz, *The Fairy-Faith in Celtic Countries*
† Jane Wilde; *Ancient Legends, Mystic Charms, and Superstitions of Ireland*
‡ Lewis Spence; *The Magic Arts in Celtic Britain*

amount of digitoxin in the stems is so acute that death can even come from accidentally drinking the water from a vase in which foxgloves have stood. This level of toxicity has given it the folk name 'dead man's bells'.

But it is not entirely without its merits. Digitoxin is still used in modern medicine to treat certain heart conditions, and historically, it was used to treat aconite poisoning and, for a brief while, epilepsy. But those who received large and repeated doses of digitalis would commonly find their eyesight affected, as the chemical also targets enzymes in the retina of the eye. This leads to a condition called xanthopsia, which causes a general haziness and a yellow tint to the vision, as well as yellow spots surrounded by coronas. It has even been speculated that Van Gogh's epileptic seizures may have been treated with digitalis, which may have caused the yellow-heavy paintings that he created in his later years, and the swirling sky in his famous *The Starry Night.* It was alluded to in his painting *Portrait of Dr. Gachet,* where he depicted his doctor holding a stem of foxglove flowers.

The use of foxglove in folk medicine required a steady hand and broad knowledge of its dangers, but it was a preferred plant of Janet Miller, a healer from the village of Dundrennan, Scotland. Her knowledge was so in demand that she travelled extensively in her parish to see to her patients, but with her fame came accusations of witchcraft, and she was eventually tried and executed in Dumfries in 1658.

FUNGI

When the moon is at the full
Mushrooms you may freely pull.
But when the moon is on the wane
Wait 'ere you think to pluck again.

Folk rhyme, Essex, England

Fungi may not technically be plants, but nonetheless worthy of mention. They are organisms in an entirely different biological kingdom, and what we commonly know as a 'mushroom' is actually the fruiting part of the fungus that allows it to reproduce. The term is typically used to denote the style of body that has a stem, cap, and gills—like the common button mushroom and toadstool—but it has become a common colloquial term used to denote all fruiting fungus bodies. Technicalities aside, it would be remiss not to discuss the wealth of lore surrounding such key inhabitants of our woodlands and hedgerows.

Ubiquitous in children's books, Christmas cards, and the Victorian revival of the fairy painting genre, the imagery of mushrooms is intrinsically linked with our vision of magic and fantasy. Early 20th Century fairy tale illustrations are rife with fairies and goblins perched upon mushrooms. Lewis Carroll's smoking caterpillar famously used one as a throne. Kitsch garden gnomes perch on them with their fishing rods and newspapers. In the world of medieval Flemish painters, toadstools often appeared in paintings of Hell. They're often a favourite metaphor of rottenness and decay for writers, appearing liberally in Shakespeare's plays, and throughout the works of classic poets such as Keats, Shelley, and Tennyson.

Perhaps it is the mysterious nature of mushrooms that makes them so interesting to us. They seem to appear overnight, with none of the careful thoughtfulness that we associate with plants; they vary hugely in size and shape and colour; and, unless you really know them, can be hugely dangerous to take home for dinner. Many of their common names are descriptive—trooping crumble caps, turkey tails, scrambled egg slime—but others are more fantastical (and forewarning), such as the destroying angel, dead man's fingers, and the corpse-finder.

Folk tales and myths about them exist across the world, such as the belief in Central America that mushrooms are in fact little umbrellas carried by woodland spirits to shelter them from the rain, which are abandoned at dawn when the spirits return home. An old Christian story proposes that mushrooms were created on the day that God and Saint Peter walked together in a field of grain. Saint Peter plucked a stalk of rye and began to chew on it, and God chastised him, telling him to spit it out, which he did. God then stated that mushrooms would grow from that grain, and they would be to feed the poor. A similar story occurs in Lithuanian folklore, in which mushrooms are considered to be the fingers of Velnias, the god of the dead, reaching out from the netherworld to feed the starving. It wasn't only in Lithuania that mushrooms have a connection to the dead; in some parts of Africa, they are regarded as symbolic of the human soul.

Many centuries earlier, ancient Egyptians believed that mushrooms grew from where lightning struck the earth, and were sent by the gods to provide immortality to those who ate them. Therefore, only Pharaohs were allowed to eat this sacred food.

Much of the fear around eating fungi comes from the difficulty in identifying many of the poisonous species, which can look misleadingly like their benign cousins. There exists a multitude of folk suggestions for sussing out the bad eggs, each one of them unfortunately not backed by science. These include placing the mushrooms amongst onions, where a poisonous specimen would turn the onions blue or brown. Similarly, parsley was supposed to yellow, or milk to coagulate, when in the presence of deadly fungi.

The Ozarks (the early English, Scots-Irish, and German settlers of the Ozark Plateau in the US) believed that mushrooms were only edible when gathered at a full moon; at any other time they would be deadly, or at least unpalatable.* This concept may have travelled across with the settlers, as a similar belief exists in the British Isles.

One concept that may have some basis in fact is the idea that a poisonous mushroom could be identified by cooking the fruit in water alongside a silver coin; or, alternately, stirring the pot with a silver spoon. If the spoon or coin turned black, the mushroom was poisonous. Silver was generally believed to be a 'pure' metal—hence the oft-repeated stance in fiction that silver bullets or knives are a good defence against werewolves and others of a supernatural ilk—and so surely something so evil as a poisonous mushroom would tarnish the metal. It may sound like mere superstition, but silver does spoil when exposed to certain gases, such as hydrogen sulphide, which may occur around the toxins of certain fungi. However the science behind the concept is not guaranteed and it is still not a recommended form of identification.

* Vance Randolph; *Ozark Magic and Folklore*

Though many mushrooms cluster harmlessly on tree stumps and boggy ground, there's one growth form in particular that's inspired tales the world over—the mushroom ring. In some countries known more colloquially as fairy rings, these formations are caused by the growth pattern of mycelium, the underground fungus that produces mushrooms. The mycelium starts at a single point, and grows outwards in a circle shape, drawing nutrients from the soil. Once those nutrients are exhausted, it pushes the circle ever-wider, occasionally causing multiple mushroom rings of gradually-increasing size. The largest—and oldest—ring recorded is said to be in Belfort, France, and is around 700 years old and 2,000 feet wide.

The formation of these rings can happen overnight, and this sudden appearance has led people over the centuries to believe that magical forces are at work. In the British Isles, it is said that these rings occur where fairies dance after a storm. However, like many places associated with the fey folk, danger will befall any mortal human who steps inside of them; a trespasser might find themselves asleep for a hundred years, or forced to dance for the entertainment of the fey until they die of exhaustion or madness. Rings aren't *always* bad luck: it's seen as fortuitous to build a

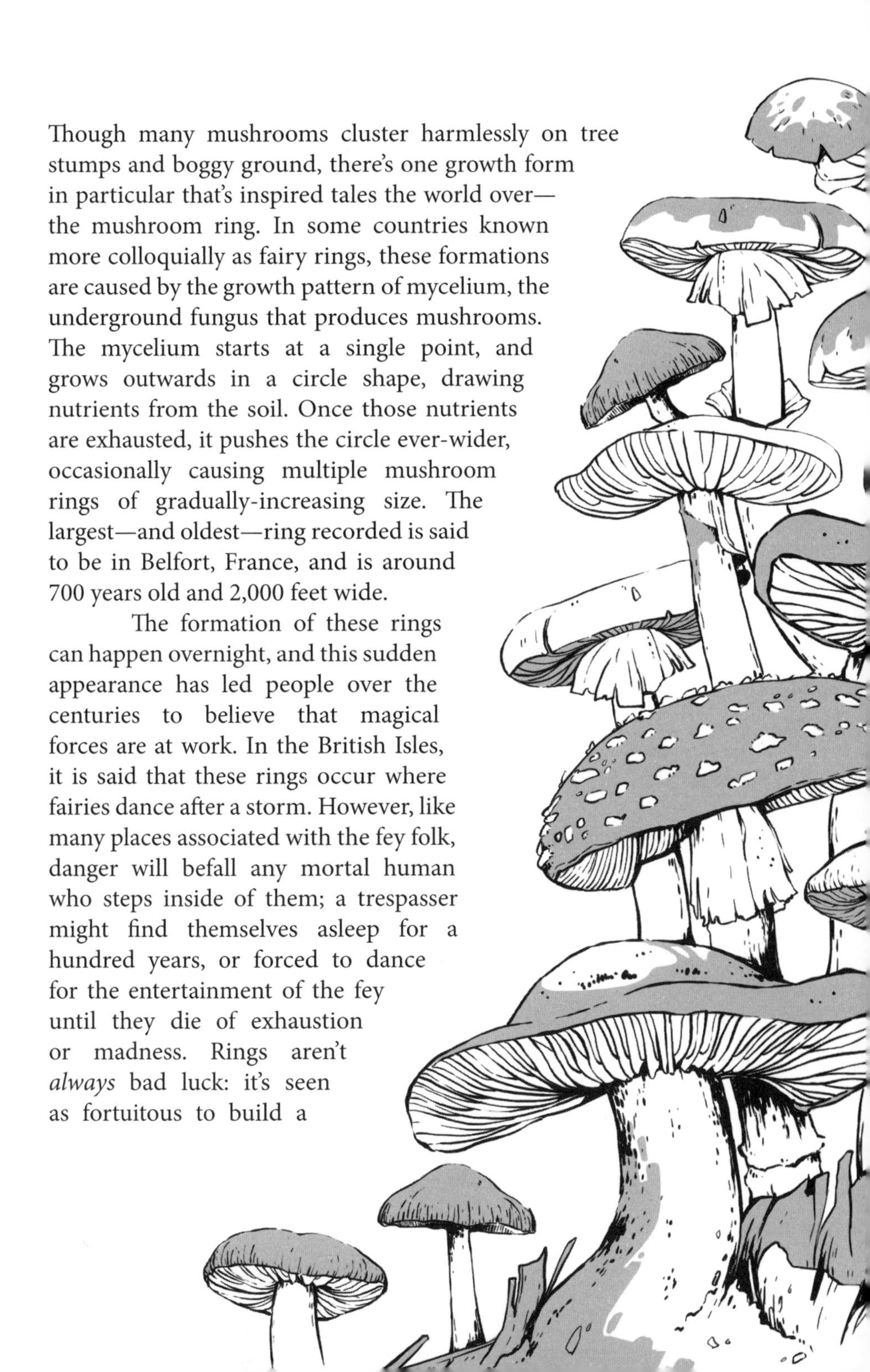

house on a field where the rings occur, and the ring itself is said to be a site of buried treasure, although to retrieve it, you'd have to ask help from the fairies themselves.

In the rest of Europe, the rings are more commonly associated with witchcraft and the Devil. In Holland, the rings are said to be where the Devil sets down his milk churn each night, and leaves a mark in the ground when he picks it up again. In France and Austria the rings are associated with dark sorcery, and are said to be guarded at night by giant toads that curse any who try to enter. In Tyrol, an historic Alpine region that is now part of northern Italy and western Austria, it was said that the rings weren't created by fairies or the Devil, but formed when a dragon would rest overnight, and scorch the earth so that only mushrooms would grow.

A similar phenomenon occurs in Namibia. These rings occur in the Namib Desert, where almost perfectly circular patches of up to 40 feet occur in the sandy grasslands, persisting for decades before suddenly disappearing overnight. Unlike their northern counterparts, these rings aren't caused by fungi; it's believed that they are the result of nests of sand termites eating the roots of the grass. These 'ghost circles', as they are known, are explained in local oral tradition as being the work of nature gods and spirits who use them as portals between the earth and the spirit world.

AMANITA FAMILY: Amanita spp.

The amanita family contains one of the most lethal natural poisons that we know of, and is thought to be responsible for ninety percent of all mushroom fatalities world-wide. Just one cap can kill an adult; it's not just the fatal dose of amatoxins that do the deed, but the fact that it can take 6-24 hours for the symptoms to show, by which point the victim is beyond medical help.

The first symptom has been reported to be, simply, a deep-seated feeling of unease. This is followed by violent cramps that seem to improve after a day or two, but ultimately lead to kidney and liver failure within a week. Though some who have eaten these mushrooms have been saved by timely transplants, most who fall foul of it do not survive the experience.

Fungi in this family include the death angel, *Amanita verna*, the death cap, *A. phalloides*, and the destroying angel, *A. virosa*.

The only member of the amanita family that is not necessarily fatal is also the one that has become the most recognised of our fungi. Fly agaric (*A. muscaria*) appears late in the year, usually between August and November, and grows typically beneath birch and spruce trees. It is instantly recognisable by its red cap and white warts. It is the 'traditional' toadstool that we're so familiar with in art and literature, and has been used throughout the ages for its hallucinogenic properties, usually as an entheogen (meaning 'summoning the divine from within') for spiritual and religious purposes.

Folklore common in Germanic and Nordic areas of Europe suggests that the name 'toadstool' comes from the manner in which toads and frogs are attracted to the fungus, and use it as a perch or shelter. The belief is certainly reflected in nomenclature across these countries; in Ireland, Wales, Germany, Norway, and The Netherlands, amongst others, it is called toad's cap, frog's cheese, frog's pouch, and toad's brain. Yet, it's rare to ever see a toad or frog interacting with these fungi, not least because their habitats are vastly different.

It's speculated that the connection comes not from toads, but from the old Breton for toad, *tousec*, which originates from the Latin *toxicum*, meaning 'poisonous'. *Kabell tousec*, the original Breton word for fly agaric, would have translated to 'the poison cap' as well as 'the toad's cap', and the variant *skabell tousec* would have meant both 'the poison stool' and 'the toad's stool'.* With the intermingling of languages within the indo-European area at the time, it's unsurprising that the translations may have at some point become crossed, to the point where folk tales now exist solely to explain the unusual name.

A little more easily explained is the common name fly agaric. *Agaric* refers to the style of mushroom with a gilled cap growing on a stalk, but the *fly* part comes from its traditional use as an insecticide; pieces of this mushroom were popularly placed in a saucer of milk or water for flies to drink from, where the poison would then kill them.

Despite being toxic enough to kill insects, fly agaric is best

* Valentina Pavlovna Wasson; *Mushrooms, Russia, and History*

known for its psychoactive effects. These are caused by two toxins, ibotenic acid and muscimol, which cause dizziness, delirium, and intoxication, and, in large enough quantities, deep sleep and then coma. To minimise the more toxic side effects, fly agaric would be processed in some way, such as drying, smoking, or being made into a drink or ointment.

In Siberia, reindeer are also that are attracted to the intoxicating effects of this fungus. It is thought that early tribesmen saw the drunken behaviour of the animals, and would slaughter them to experience the same effects from eating the meat. Another, less appealing form of preparation involved drinking the urine of reindeer that had feasted on the mushrooms; the toxins remain active within the water, but without the dangers of overdose.

It is a more recent suggestion that early development of the legend of Santa Claus may have come from the same part of the world. Recorded amongst the Koryak people of Kamchatka in Siberia, but existing in similar forms amongst other peoples of the north-eastern areas, is a midwinter festival that sees the local shaman enter the yurt through the smoke hole, carrying gifts of dried fly agaric. He ingests the mushrooms during the ceremony, and those participating drink his urine to partake of the fungus's entheogenic effects. It may sound unappetising, but the shaman will have fasted for several days before the ceremony, at which point his urine consists of mostly water and the unaltered hallucinogenic compounds. The point of the ceremony is to spiritually journey to the tree of life—a large pine that grows by the North Star—to find the answer to all of the village's problems from that year. After the ceremony, the shaman leaves in the same manner, up the birch lodgepole in the centre of the yurt and out the smoke hole. The tribespeople believe that he can fly, or enters and leaves with the aid of the flying reindeer who share the mushrooms with him.

In parts of northern and eastern Europe, fly agaric is also called 'Raven's Bread', in connection with the two ravens, Huginn and Muninn, who travel with the Norse god Odin. This dates back to Middle Ages skaldic poetry, where the fungus is referred to as *Munins tugga*—food of Munin. *Munins tugga* or *verõr hrafns* ('raven's food') was also used in skaldic poetry to symbolise corpses, and the redness of blood (or fly agaric) on snow. The mushroom was believed to grow where foam fell from the mouth of Sleipnir, Odin's horse, during the course of the wild hunt at the winter solstice. From this foam, the earth would become impregnated with fly agaric and would then begin to grow nine months later in the autumn.

Back in Kamchatka, the raven is a sacred animal and cultural hero for the people who live there. In the mythology of the Korjaken people, the fly agaric grew from the spittle of the Creator when he spat on earth, and was then eaten by the Great Raven. When Raven realised its clairvoyant properties, he declared that the fly agaric must forever grow on the earth, so that humans might see what it has to show them.

Beyond the spiritual, there are suggestions in multiple historic texts that fly agaric was utilised by the Vikings in their famous berserker rages, during which their warriors were renowned for their uncontrolled rage and fearlessness. Ingestion of amanita is known to inhibit the body's fear response and startle reflex, which would have been invaluable in battle.

Another famous mythical figure thought to have used amanita for this purpose is Cuchulainn, the famous warrior demigod of Irish and Scottish legend. Cuchulainn was famous for being able to enter *riastradh,* the 'battle spasm', which not only allowed warriors to become frenzied but also included feats of huge strength, an unstoppable desire to kill everyone in sight, physical and facial distortions, and a great heat throughout the body. All of these symptoms are typical of amanita ingestion. The last in particular—the 'fire in the head', as it is known—refers to a common symptom of amanita, where heat rushes into the face and brain. Ingestion of this particular fungus might also explain the great 'wasting sickness' that he suffered after such bouts, where he would fall into a great depression and sleep for long periods, as discussed in *Serglighe Con Culainn,* or 'The Sickness of Cuchulainn', a narrative from the Irish *Ulster Cycle.*

It is thought that the appearance of the fly agaric may have inspired myths of the redcap, a malevolent goblin-type creature found in

Border folklore. This creature is said to inhabit ruins and abandoned castles across the Anglo-Scottish border, particularly those that were the sites of murders or battles. Appearing as a short and fierce old man, the redcap has long teeth and prominent claws, and wears on his head a scarlet-coloured cap. Should anyone enter his lair, he stones them to death, and soaks his hat in their blood. A common belief was that the redcap haunted these areas as the Border castles were built by Picts who bathed the foundation stones in human blood.

Further south in Cornwall, 'redcap' is used as a more general term to refer to benevolent trooping fairies, which were known for wearing green jackets, scarlet hats, and a white owl's feather.

FALSE MOREL: Gyromitra esculenta

The false morel, as its name implies, looks remarkably like the edible morel. Despite the specific epithet *esculenta*, which means 'delicious', it should not be considered edible. It's a popular delicacy in Scandinavia, Finland, and Poland, but is banned in other European countries and must be sold with warnings regarding its proper preparation. Though it can be eaten safely in the hands of someone familiar with cooking it, boiling the mushroom can cause the toxins to vaporize, which can cause sickness just from inhaling the steam.

Its toxin, monomethyl hydrazine, causes vomiting, dizziness, diarrhoea and eventually death. Though not always fatal, it is thought to

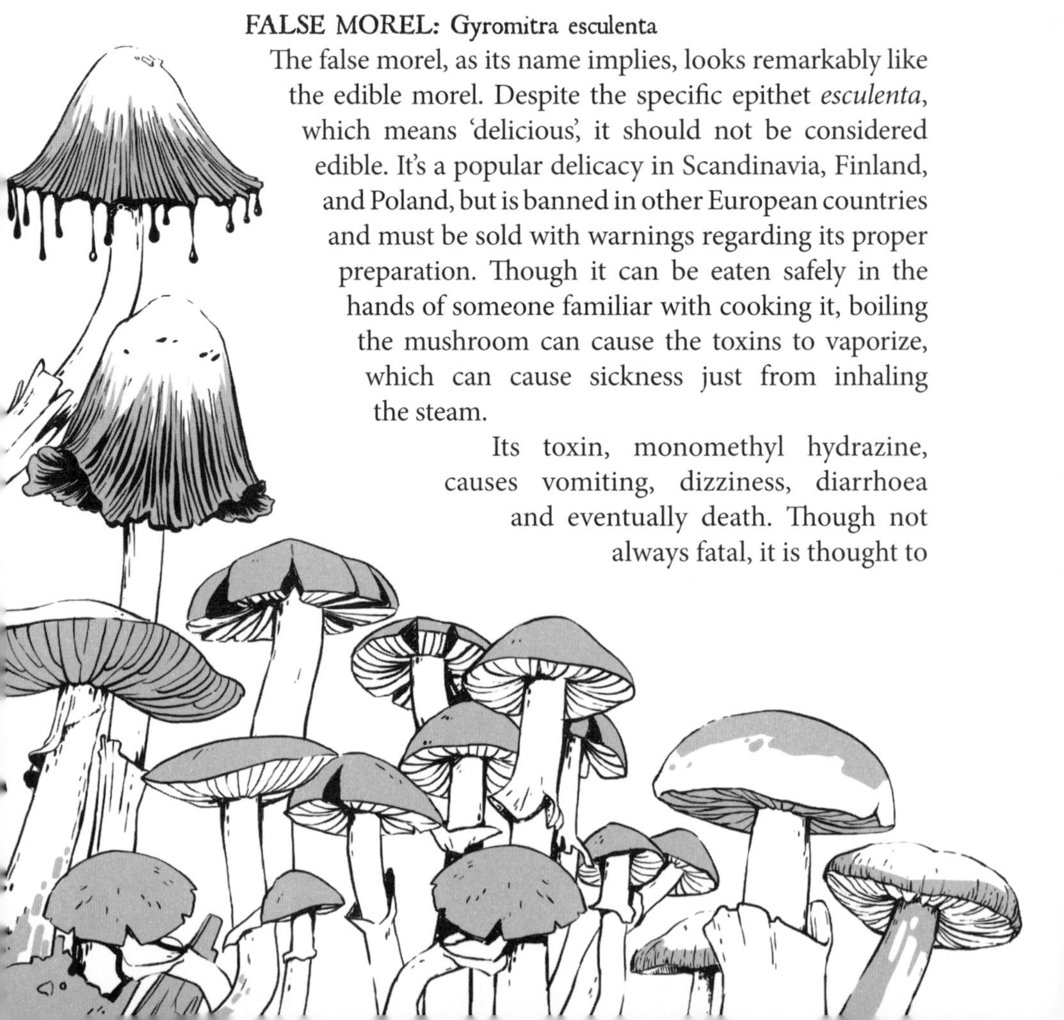

be responsible for almost a quarter of mushroom fatalities in Europe each year.

FROSTED FIBRE CAP: *Inocybe maculate*

The frosted fibre cap contains muscarine, which is deadly both via ingestion and inhalation. It causes a slowed pulse, sweating, and lack of coordination caused by a disturbance of the autonomic nervous system. Death is eventually caused by respiratory failure.

GOLDEN JELLY FUNGUS: *Tremella mesenterica*

The golden jelly fungus is also known as the yellow trembler, and is a common fungus across the globe. Although it is edible and non-toxic (although flavourless), in Sweden it is called Witch's- or Troll's Butter, and is said to be capable of bestowing curses. To curse an intended victim, all that was required was to throw the fungus at them to cause a turn of bad fortune.

However, the victim was able to turn the curse back on the person who had placed it. By striking the fungus, the curse would be lifted; if it was struck with a blunt weapon then the curser would be maimed, and if a sharp weapon, the curser would die.*

INKY CAP: *Coprinus atramentarius*

The inky cap is a small, white mushroom that turns black when mature, and begins to 'melt' in long, inky droplets. A common fungus across the Northern Hemisphere, it is entirely edible, unless ingested alongside alcohol—at which point it becomes poisonous, hence another common name, tippler's bane. This effect is caused by the active compound in the mushroom, called coprine, which blocks the enzyme in the body that usually breaks down the parts of alcohol that cause a hangover. As a result, the symptoms of inky cap poisoning can include facial reddening, nausea, vomiting, malaise, agitation, palpitations and tingling in limbs. Though most recover from this within a few hours, it has been known to lead to cardiac arrest.

* Johannes Björn Gårdbäck; *Trolldom: Spells and Methods of the Norse Folk Magic Tradition*

SATAN'S BOLETE: Rubroboletus satanas
Satan's bolete, also known as the bloody boletus, is a member of the porcini group but, unlike its cousins, isn't edible. Reports of poisoning from this fungus are rare, as its off-putting appearance has no doubt discouraged experimentation from mushroom hunters. Squat and red-bodied, it smells of rotting flesh, which grows stronger as it ages, and the flesh turns blue when cut or bruised. Symptoms of ingestion include nausea and vomiting.

These mushrooms, which can grow up to 30cm wide, were first named and recorded in 1831 by German mycologist Harald Othmar Lenz. Though the name may relate to the colour of the mushroom and the red coat that Satan is said to wear, it may have been given out of spite by Lenz, who claims to have been made ill by its 'evil emanations' when studying it.*

SCARLET ELF CUP: Sarcoscypha coccinea
The scarlet elf cup is a striking-looking fungus that can be found growing on dead wood throughout early spring. The bright red cups rarely grow any larger than an inch, and are said to be the drinking vessels from which elves and fairies drink the morning dew. This folk belief is reflected in the scientific name: *sarco* means 'flesh', and *skyphos* 'drinking bowl'.

Though technically edible, it is not a popular food due to its bland flavour and tough texture. However, the Oneida and other first nations tribes of the Iroquois area have historically used it to promote healing and stop bleeding, applying it to the navels of newborns or over wounds that are otherwise slow to heal.

SPIT DEVIL: Russula emetica
The spit devil, also known as the sickener, is a peppery-tasting fungus that grows mostly beneath conifer trees. As suggested by its name, eating it can cause serious gastrointestinal distress, which can lead to hours of discomfort, though is rarely fatal. The toxins can be removed by parboiling or pickling, and it's a popular food in Russia and Eastern Europe.

* Peter Marren; *Mushrooms: The Natural and Human World of British Fungi*

GAS PLANT: Dictamnus albus

And wood nymphs in white array
lovely, beautiful, take up the song—
softly treading the verdant grass
'Til they reach the hero and sit down.
One binds his wound with herbs
another splashes him with water
a third hastens to kiss his mouth
as he gazes at her—lovely, smiling.

Hristo Botev, *Hadzhi Dimitar*

With its bushy leaves and tall, pink-flowered spikes, the *Dictamnus* is a striking plant that hides an unexpected secret. Beneath its delicate blossoms and lemon-scented leaves, it produces a volatile oil prone to starting fires, and when exposed to a match it is engulfed in flame, with no harm done to the plant itself.

This ability to set itself alight comes from the isoprene oil that forms on the leaves of the plant, which evaporates at low temperatures and forms a highly flammable atmosphere around it. This flammability is thought to simply be a side effect of the oil, which is designed to protect against heat stress. It is also this oil that causes the distinctive lemon scent of the plant, which reveals its unlikely relatives: it's a member of the citrus family.

The gas plant—or fraxinella ('little ash') as it's also known, due to the shape of its leaves—is said to grow near the water sources favoured by *samodiva* and *samovila*, playful Bulgarian woodland spirits who can be either beneficent or maleficent to those who come across them. Their names betray their nature—the prefix *samo-* means 'self', and *diva* suggests 'wild' or 'raging', whereas *vila* means 'spinning', in the manner of a tornado or wild wind. These nymphs dress in white and wear wreaths of fraxinella leaves, and are said to transform into wolves, or ride through the woods on white deer and bears.*

Though mostly benevolent, when angered they show their natural affinity for fire—they are said to be able to transform into monstrous birds that throw flames, and can also bring about drought, or make cattle die of high fever.†

* Mihail Arnaudov; *Snapshots of Bulgarian Folklore*
† Jan Máchal; *The Mythology of all Races. III, Celtic and Slavic Mythology*

GHOST PLANT: Monotropa uniflora

It is the weirdest flower that grows, so palpably ghastly that we feel almost a cheerful satisfaction in the perfection of its performance & our own responsive thrill, just as we do in a good ghost story.

Alice Morse Earle

The ghost plant is an unusual flower which, despite its striking appearance, is almost too easy to miss among the loam. Although it grows in midsummer, the clusters of these plants are spectrally pale, and are cold and clammy to the touch. Furthermore, touching them produces a sticky liquid and causes instant bruising, which turns black and then causes the plant to appear to 'melt'. Also known as ghost pipe and corpse plant, it is found across North America, Japan, and the Himalayas.

This flower cannot photosynthesise, and depends on very specific trees and fungi for nutrients. The roots of the plant have formed a symbiotic relationship with another parasitic plant: a fungus that relies on leaf mould on the ground, which connects the ghost plant to the roots of nearby conifers so that they can both feed off the sugars that come from the tree. Without the complex and unique combination that is the conifers, sufficient leaf litter, and the fungus, this plant simply could not survive.

The Coast- and Straits Salish groups (a group of indigenous peoples on the Pacific Northwest coast) associate this parasitic flower with wolves, and its local name means 'wolf's urine', as it is believed to grow wherever a wolf marks its territory. The connection likely comes from the strong ammonia scent of the plant, and its growing habitat in sites popular with wolves.

Most likely due to its spectral appearance, in Europe it has become associated with the afterlife and ghosts. One German folk remedy claims that this plant can heal a heart that has been broken through the death of a loved one.

As an important note, in the past decade this plant has moved from being listed as common to endangered, most likely due to increased awareness of its curious existence and discussion about its medicinal properties. It does not survive picking or transplanting due to the complex manner in which it exists, and populations have been completely decimated by curious admirers. If you encounter it, it is important to remember that many things are more beautiful left exactly where they are.

GIANT HOGWEED:

Heracleum mantegazzianum

And I saw it again, in a far
northern land—
Not a pansy, not purple and white;
Yet in beauteous guise
Did this poison-plant rise,
Fair and fatal again to my sight.
And men longed for her kiss and her odorous breath
When no friend was beside them to tell
That to kiss was to die,
That her truth was a lie,
And her beauty a soul-killing spell.

John Boyle O'Reilly,
The Poison Flower

Giant hogweed is a highly invasive weed that is thought to have originated in the Caucasus region of Eurasia, but which has rapidly spread across all parts of the globe. Its initial introduction was helped by early botanists who, taken by its white flowers and large size (some specimens can grow up to 18 feet tall), brought it back to their home countries as an ornamental plant for large gardens. It is recorded as being introduced to England in the late 1800s as a gift to the Royal Gardens at Kew. It didn't take the plant long to escape and it now grows freely across the British Isles.

A relative of some more innocent and familiar species, such as parsley, carrots, and coriander, it is also related to a number of other, more deadly suspects mentioned in this book, such as hemlock, fool's parsley and water dropwort.

And, like these troublemakers, giant hogweed is a pest not just because of its rapid growth habit, but also because coming into contact with it can cause serious damage to the skin. The juice of the plant is phototoxic, and can cause blisters, burns, and even blindness. The sap, when coming into contact with the body, stops the skin from being able to protect itself from sunlight, even on overcast days. The initial photosensitivity—which can lead to skin inflammation and burning—can last for several days, but can also re-emerge for years afterwards; some sufferers have reported the burns returning for twenty or thirty years after exposure. In some severe cases, the sap can cause permanent changes to skin colour.

In a Swiss study of plant poisoning reports placed over 29 years,* the number of giant hogweed incidents with 'serious' consequences made it the second most dangerous plant in the country, second only to deadly nightshade.

* Jaspersen-Shib; *Serious plant poisonings in Switzerland 1966-1994. Case analysis from the Swiss Toxicology Information Center*

GRASSES

Low musick, thin as winds that lyre the grass,
Smiting thro' red roots harpings; and the sound
Of elfin revels when the wild dews glass
Globes of concentric beauty on the ground;
For showery clouds o'er tepid nights that pass
The prayer in harebells and faint foxgloves crowned.

Madison Julius Cawein, *Disenchantment of Death*

Scant few places in this world are without the sight of grass. So mundane that they are often overlooked, grasses are everywhere—in lawns, fields, woodlands, and appearing as river rushes, bamboo, sugar cane, and cereal plants, to name but a few. But for all the good they provide, some species of this large and varied family are less inclined to be friendly.

HUNGRY GRASS

In Irish mythology, *féar gortach*, or hungry grass (also known as fairy grass) is a patch of cursed grass that dooms anyone who walks upon it to a perpetual and insatiable hunger. Though there is no set story as to how such an area comes to be cursed, it has been suggested in some stories that the grass is planted by fairies to capture wary passers-by,* whereas others suggest the curse arises due to the proximity of an unshriven corpse.†

Typically, when a person died their corpse would undergo a 'watch' for a number of days before its ultimate burial. In this time, friends and family would keep the coffin company, remembering the deceased person fondly and eating and drinking around it to ensure that the corpse did not rise for loneliness. However, if the corpse was at any point left alone, it might rise and wander away, and thus become a *féar gorta*. Similar in name to the hungry grass, this gaunt corpse would travel and ask people for food and money. Those who give freely would

* William Carleton; *Traits and Stories of the Irish Peasantry, Volume III*
† Steenie Harvey; *Twilight Places: Ireland's Enduring Fairy Lore*

be rewarded with good fortune, whereas those who do not would be punished with poverty. It is thought that hungry grass occurs wherever this corpse has walked, cursing the ground beneath it.

If you ever do come across a patch of hungry grass, a fix for the curse is simple: it's suggested that all you need to safely cross is to always carry with you some food and some beer to consume along the way. A good suggestion in any situation, really!

COGON GRASS: Imperata cylindrical

Native to China, cogon grass—also known as Japanese blood grass—is an invasive plant that travelled across to Japan in the 1700s and began to invade the US in the 1940s. Growing up to ten feet tall in places, the edge of each blade of grass is tipped with tiny silica crystals that are as sharp as a knife, and even the roots are barbed so that they can cut through the roots of other plants that compete for the same resources.

This fire-adapted species is highly flammable, and cause fires that burn hotter and brighter than normal blazes, due to the density of the plant's growth. These fires are hot enough to kill competing plants, even trees, and once the competition is gone the earth is barren enough for young shoots to grow up from the grass's underground rhizome network.

In relation to the way that cogon fires burn and then die rapidly, in the Philippines, someone who is called *ningas cogon* ('a cogon brush fire') is a person who procrastinates, specifically someone who shows a great interest in a project that dwindles quickly.

JOHNSON GRASS: Sorghum halepense

Johnson grass is a tall grass that is invasive across the US. It is most dangerous to cattle, which might graze upon the young shoots while they are still small; the shoots alone have enough cyanide to kill a horse, and causes anxiety and convulsions before eventual cardiac arrest.

Another member of the sorghum family, broom-corn (*Sorghum bicolor)*, was said to be the chosen weapon of witches in 16th Century Friuli, a region of north-east Italy. These malevolent witches (called *malandanti,* 'bad walkers') did battle with sorghum stalks against the *benandanti* ('good walkers'), who carried fennel. These battles were done at night, in a dream world away from their physical bodies.* This form of dream-walking is similar to the Corsican *mazzeri,* who do battle with stems of asphodel, and the concepts may have originated from the same practice. During the witch trials of the 16th Century the benandanti were accused of being witches themselves, and the name benandante became synonymous with *stregha,* the original Friulian word for witch.

SWEET GRASS: Hierochloe odorata
Sweet grass is a hardy plant that's able to grow even in the Arctic Circle. Though each blade can grow up to seven feet long, it does not have the rigid stems needed to grow tall and therefore grows out horizontally. Popular across Europe for its sweet scent and flavour, it is used in France to flavour sweets, tobacco, and drinks, in Russia as a tea, and in Germany it is strewn across church doorways on holy days. This tradition provides the roots of its scientific name: *hierochloe* is Greek for 'holy grass', while *odorata* means 'fragrant'.

* Carlo Ginzburg; *The Night Battles: Witchcraft and Agrarian Cults in the Sixteenth and Seventeenth Centuries*

In Poland, where it is known as bison grass, it is an ingredient in *Zubrowka,* a traditional vodka. Though the grass is included in the drink to add its distinctive sweet flavour, the sweetness is provided by coumarin, which also thins the blood and relaxes the body. Though the drink is still available across Europe, it has been banned in the US since 1978.

GREAT MULLEIN: Verbascum thapsus

I am too near, too dear a thing for you,
A flower of mullein in a crack of wall,
The villagers half-see, or not at all,
Part of the weather, like the wind or dew.

Lizette Woodworth Reese, *A Flower of Mullein*

Great Mullein is a tall plant that grows long spikes of flowers from a central rosette of leaves. Due to its fast-growing nature and the long life of its seeds, it has come to be seen as something of a weed and pest in many areas.

Despite its problematic growth habits, it has its uses: the flowers can be used to produce a bright yellow or green dye, and in early Roman times, it was known as *Candela regia* or *Candelaria*, as the stalks would be dried and dipped in suet to burn as torches at funerals.* This use wasn't restricted to the Romans; in northern Europe it was colloquially called hedge-taper, the name which became convoluted to hag-taper during the period of the Europe-wide witch trials in the 16th and 17th Centuries, casting suspicion on any who allowed it to grow near their homes. The plant is ideal for use in fire-making; the wool from the leaves and stems can be plucked off for tinder, the stringy fibre of the stems makes ideal wicks, and the entire flowering stem of the plant can be dried and use as a torch. This use made it an ideal guard against darker powers. In Europe and Asia both it was said to

* John Parkinson, *Theatrum Botanicum*

have the power of driving away evil spirits, and in India it would be burned as a ward against evil spirits and magic. The Greeks ascribed it the same power; in Homer's *Odyssey*, Odysseus carried this plant to protect himself from Circe's wiles while staying on Aeaea.

Careful handling of this plant is advised, as the hairy leaves and caustic juice can irritate the skin, and the entire plant contains rotenone, a toxin chemically related to the venom found in the skin of toads. Ingestion can cause tinnitus, vertigo, and thirst, as well as suffocation, swelling of the tongue and throat, and slowing of the heart. In severe cases the heart can be slowed to such an extent that it causes cardiac arrest.

HELLEBORE, BLACK: Helleborus officinalis

Nor will the lantern'd fisherman at eve
Launch on that water by the witches' tower,
Where Hellebore and Hemlock seem to weave
Round its dark vaults a melancholy bower,
For spirits of the dead at night's enchanted hour.

Thomas Campbell, *Lines on the Grave of a Suicide*

This well-known plant, common wild across Europe, belongs to the Ranunculus order alongside its toxic cousins, buttercups. Those seeing the black hellebore for the first time might be surprised by its pink and green flowers, but the name comes not from the colour of the petals, but the roots. Many older herbals make a distinction between black and white hellebore, but in the current day 'white' hellebore does not exist; it has since been identified as *Veratrum viride*, the false hellebore, which is even more poisonous, but unrelated. Black hellebore is also commonly known as the Christmas Rose due to its time of flowering, however it is no relation to the rose family.

The name *Helleborus* comes from *elein*, meaning 'to cause death', and *bora*, 'food'. As suggested by this, the plant—and all other species in this family—are highly toxic, and have a long history of use as pesticides and instruments of warfare. All parts of the plant are poisonous, and can cause skin irritation if handled incorrectly, or if eaten, can cause burning of the mouth and throat, vomiting, and disruption to the nervous system. Even just smelling the plant can cause burning to the nasal passages, and there are records of death coming within eight hours to someone who has swallowed even one ounce of water in which the roots have been soaked.*

These properties were taken full advantage of by the Greeks, who used the plant as an early form of chemical warfare. An account from Pausanias, an early Greek geographer, tells of an attack in 595BC by Athenian troops on the town of Cirrha, which had been increasing tolls for pilgrims to the holy sanctuary of Pytho (later called Delphi). During the siege, the canal that supplied the town with water was blocked in an attempt to force the Cirrhaeans to surrender. When this failed, Solon—the commander of the attacking Athenian army—allowed the water to resume its course, but first soaked in it great bundles of hellebore. The thirsty troops within the city became so sick from drinking the tainted water that they were too weak to hold the walls, and thus the Athenians succeeded in their siege.

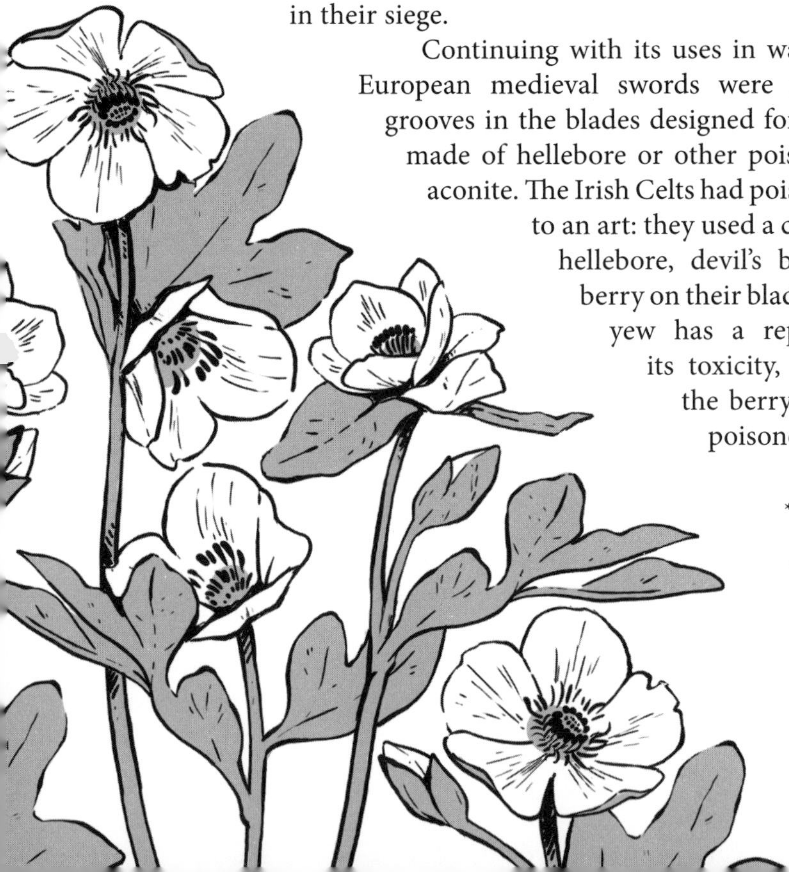

Continuing with its uses in warfare, many European medieval swords were forged with grooves in the blades designed for fatal pastes made of hellebore or other poisons such as aconite. The Irish Celts had poisoning down to an art: they used a compound of hellebore, devil's bit, and yew berry on their blades.† Though yew has a reputation for its toxicity, the flesh of the berry itself is not poisonous, and

* William Thomas Fernie; *Herbal Simples Approved for Modern Uses of Cure*

† Robert Graves; *The White Goddess*

was likely used for its stickiness in binding the combination of other plants together.

By the 16th Century, however, hellebore had little use in conflict, but its toxicity was still valuable to many. The author Leonard Mascall describes in his 1590 *A Booke of Engines and Traps* how to utilise hellebore to rid a house of unwanted pests: 'Take the powder of Ellebore, otherwise called micing powder, and mire it with barley meale. Then put to honny and make a paste thereof, then bake it, or leeth it, or frie it, and it will kill those mice that eates thereof.'

The root has been noted for centuries as a cure for insanity, though the origins and veracity of this are dubious and hard to prove. It may have started with the Greek myth of Melampus, who was said to have cured, with hellebore, the madness of the daughters of King Proteus of Argos—a story which herbalists of the 15- and 1600s (who often wrote of myth and reality as one and the same) took as license to spread the belief further still. This belief also gave rise to the Roman proverb *Naviget Anticyram:* 'take a voyage to Anticyra', the Grecian island where hellebore grows abundantly. This would often be said to a man who was seen to have lost all reason.*

As with many poisonous plants, hellebore historically became attributed with magical abilities, and was favoured by witches and doctors alike. It was said to have the capacity to alter or change the nature of another plant if used as a fertiliser or grafted directly onto the stem of the plant in question. Through this technique, it was believed that its poisonous qualities could be magically transferred to plants that were typically benign. These plants would then be ideal for blighting, or blasting—an early English form of cursing. In France, it was thought to have the power to alter perception of the space around it; folk tales still circulate of army-conscripted sorcerers who were able to move about unseen through the enemy by surrounding themselves with a cloud of powdered hellebore.†

It was also used by doctors to 'cure' illnesses that were believed to have been inflicted by witchcraft. It was said to be efficient in curing

* William Thomas Fernie; *Herbal Simples Approved for Modern Uses of Cure*
† Maude Grieve; *A Modern Herbal*

magic-borne deafness, and in the 1600s was thought to cure those possessed by the Devil, leading to a period of time where it was known as *fuga daemonum*, 'Devil's flight'.* A number of plants with the same attributes, such as St John's Wort, were also given this name around the same period.

HELLEBORE, FALSE: Veratrum spp.

In the east the day was reddening,
When the warriors pass'd;
In the west the night was deadening,
As they looked their last;
As they looked their last on him—
He, their comrade—their commander
He, the earth's adored—
He, the godlike Alexander!
Who can wield his sword?

Letitia Elizabeth Landon, *The Death-Bed of Alexander the Great*

For a long time, and in many early herbals, *Veratrum* was referred to as white hellebore and believed to be a true part of the *Helleborus* family, despite the lack of visual similarities between the two plants. Now known to be unrelated, the false hellebore sits in a genus of its own, and just as well; it's even more deadly than the true hellebore.

Where the true hellebore is rarely fatal, poisoning by false hellebore is fast and deadly. It causes tinnitus, vertigo, stupor, and an unbearable thirst, followed by a sensation of suffocating and violent vomiting. The heart is then slowed, leading to seizures and cardiac arrest. However, the toxins in this plant are only produced during active growth; in the winter months, most of the toxins degrade and it is in this period that the roots can be gathered for medicinal use, a practice observed by some western American Indian peoples such as

* Richard Folkard; *Plant Lore, Legends, and Lyrics*

the Blackfoot tribe. The root is also gathered by the Blackfoot in the spring and summer, when it is at its most toxic, and reserved for suicidal purposes by those suffering otherwise incurable diseases.*

First Nations people have other uses for the root. In 1636, John Josselyn noted that some tribes held an ordeal by which to choose their next chief; those who chose to undergo it would ingest the root of the plant, and whoever could last the longest without vomiting was declared to be the strongest of them.† The Salish peoples of Vancouver Island believe that its poisonous properties extend even further; when at sea, they carry it as a charm to kill sea monsters that might otherwise be impervious to attack.‡

However, the plant remains a highly poisonous one, and contact with it is discouraged. The most famous victim of this poison is Alexander the Great (or so it has been speculated), who died at the age of 32 in 323BC after having conquered most of the known civilized world at the time. Reports of his declining health were recorded by Plutarch and Diodorus over the two weeks preceding his eventual death, and the symptoms are consistent with hellebore poisoning.

* Alex Johnston; *Blackfoot Indian utilization of the flora of the north-western Great Plains. Economic Botany. Vol 24*

† John Josselyn; *New-England's Rarities Discovered in Birds, Beasts, Fishes, Serpents, and Plants of That Country*

‡ Nancy Turner and Marcus Bell; *The Ethnobotany of the Coast Salish Indians of Vancouver Island*

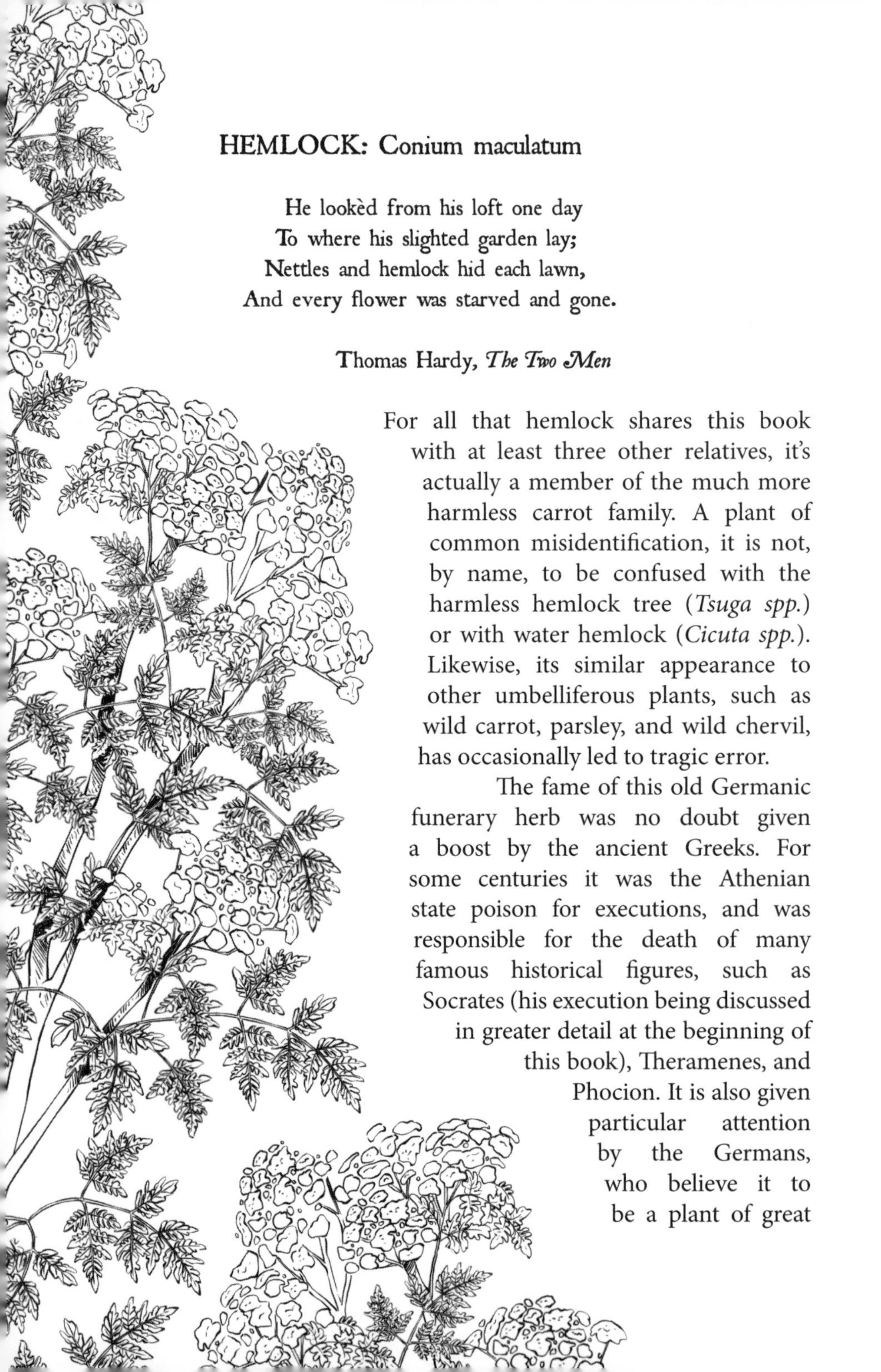

HEMLOCK: Conium maculatum

He lookèd from his loft one day
To where his slighted garden lay;
Nettles and hemlock hid each lawn,
And every flower was starved and gone.

Thomas Hardy, *The Two Men*

For all that hemlock shares this book with at least three other relatives, it's actually a member of the much more harmless carrot family. A plant of common misidentification, it is not, by name, to be confused with the harmless hemlock tree (*Tsuga spp.*) or with water hemlock (*Cicuta spp.*). Likewise, its similar appearance to other umbelliferous plants, such as wild carrot, parsley, and wild chervil, has occasionally led to tragic error.

The fame of this old Germanic funerary herb was no doubt given a boost by the ancient Greeks. For some centuries it was the Athenian state poison for executions, and was responsible for the death of many famous historical figures, such as Socrates (his execution being discussed in greater detail at the beginning of this book), Theramenes, and Phocion. It is also given particular attention by the Germans, who believe it to be a plant of great hatred, and claim that it harbours special resentment for other plants that are more favoured by the people, such as rue, for which its distaste is so strong that it will grow nowhere near it.

Some historical accounts remark on death by hemlock to be violent, and characterised by choking and convulsions. This is a good example of why it helps to be aware of mistranslations or misidentifications; convulsions are a symptom of water hemlock, the *Cicuta* family, and not *C. maculatum*. True hemlock was chosen by the Greeks for to its careful, slow nature; death by hemlock takes several hours, and is so gradual that many doses were not immediately fatal and may have needed to be topped up during the execution. This was the case in the execution of Phocion, an Athenian statesman known for his honesty. According to an account of proceedings, the initial dose of hemlock wasn't enough to finish the deed, and, taking advantage of his position as the only one qualified to administer more, the executioner refused to prepare a second dose unless he was paid an extra 12 drachmae.

Under certain conditions, suicide was seen to be noble in the time of the ancient Greek empire, and in these cases use of it was often sanctioned. On the Aegean island of Cea (modern-day Kos), inhabitants, upon reaching a certain age or feeling as though they had achieved all they would in life, would take hemlock so as not to be a burden in their old age. An account from the philosopher Michel de Montaigne, tells of one such party that was thrown in the presence of Sextus Pompeius, a 1st Century Roman General:

Sextus Pompeius, in his expedition into Asia, touched at the isle of Cea in Negropont: it happened whilst he was there that a woman of great quality, having given an account to her citizens why she was resolved to put an end to her life, invited Pompeius to her death, an invitation that he accepted ... She had passed the age of four score and ten in a very happy state, both of body and mind; being then laid upon her bed, better dressed than ordinary and leaning upon her elbow, "For my part," said she, "having always experienced the smiles of fortune, for fear lest the desire of living too long may make me see a contrary face, I am going, by a happy end, to dismiss the remains of my soul, leaving behind two daughters of my body and a legion of nephews"; which having said, she boldly took the bowl that contained the poison, and having made her vows and prayers to Mercury to conduct her to some happy abode in the other world, she roundly swallowed the mortal poison. This being done, she entertained the company with the progress of its operation, and how the cold by degrees seized the several parts of her body one after another, till having in the end told them it began to seize upon her heart and bowels, she called her daughters to do the last office and close her eyes.

Michel de Montaigne, *Essais*, 1580

A similar account from Valerius Maximus, a writer who was a close friend of Sextus Pompeius, speaks of a law in Marseilles that existed around the same period. At that time, it was unlawful to take your own life without regulation; however, those who wished to do so (usually out of a desire not to be a burden on younger generations) could speak their case before the senate and give the reasons for their wish. If the senate agreed that the motives were lawful and not forced by others, they would allow the person access to a poison prepared from hemlock.*

HEMLOCK WATER DROPWORT: Oenanthe crocata

The nature is thought to kill laughing, but without doubt the thing is clean contrary, for it causeth such convulsions, crampe, and wringings of the mouth and daws, that it hath seemed to some that the parties have died laughing, whereas in truth they have died in great torment.

John Gerard, *Great Herball*

Found in ditches, riverbanks, streams and other damp environments, Hemlock water dropwart is made distinct by its height (up to five feet tall) and the white umbelliferous flowers that show in July. Beneath the

* Valerius Maximus; *Book II*

soil, the plant grows from clusters of long, white or pink tubers—the reason it is colloquially known in some areas as 'dead man's fingers'.

The name *Oenanthe* comes from the Greek *oinos* and *anthos*, meaning 'wine flower', due to the fruity scent of the flowers when blooming. A record from Captain J Palliser in 1863 (Palliser led the British North American Exploring Expedition into what would later become western Canada) mentions an incident regarding this plant that occurred while the expedition camped near a swamp not far from the Pembina River in Alberta. Amongst his companions were a number of trackers born to Iroquois Indians and French settlers, and knowledgeable about the folklore of both cultures. They called it *carrot à moreau*, and when the camp was plagued at night by the bodiless sound of muttering in the swamps, they claimed that the noise must come from the plant, 'in consequence of its poisonous and miraculous attributes'. Unsettled by these claims, a number of men set out into the swamp to discover the truth, only to have the noise stop whenever they grew near to its Source. The trackers agreed that this was a known quality of the plant, and that it would grow silent upon approach to better hide itself. After some time searching, the culprit was finally apprehended—a small frog going about its night-time business, quite unaware of the disturbance it had caused.

The source of ghostly mutterings or not, the fact is that the water dropwort is remarkably poisonous. It is one of the deadliest European plants, and a strong contender with native species in other countries. With a stem that resembles celery and roots that resemble its benign, edible cousins, it is no surprise that this plant is responsible for a number of poisonings, both human and animal. One root is more than sufficient to kill a large cow, and less than this has brought about the end to ill-adventuring humans.

The stem and roots contain large amounts of oenanthotoxin, which can cause convulsions, seizures, kidney failure, and eventually respiratory and cardiac distress. It also can lead to paralysis of the voice box, resulting in long-term muteness. This symptom, giving it the folk name 'dead tongue', was noted by the Anglo-Irish herbalist Threkeld to have been responsible for the hospitalisation of eight boys in the early 1700s; five of them died before the next morning, not one of them having spoken a word since eating the plant.*

Water dropwort is also thought to be responsible for the origins of the word 'sardonic', meaning 'to be grimly mocking'. In early Greco-Roman literature is recorded a plant known as the sardonic herb, which has now been identified as a member of the *Oenanthe* family. The convulsions caused by ingestion of the plant can spread as far as the facial muscles, causing a condition known as *risus sardonicus*, 'the scornful laughter'. *Risus sardonicus* causes the eyes to bulge, the eyebrows to rise excessively, and the lips and mouth to retract dramatically, causing the impression that the victim has died of laughter. This condition has been noted not only in victims of poisoning (strychnine causes similar symptoms) but also in those suffering from tetanus. It was Homer who first coined the word 'sardonic' after learning that the Punic people in Sardinia used this plant to poison elderly people or criminals before throwing them from a cliff. The plant was ideal for this ritual, as the Punics believed that death was simply the beginning of a new life, and should be greeted with a smile.

* Caleb Threlkeld; *Synopsis Stirpium Hibernicarum*

HEMP: Cannabis sativa

Shudd'ring at the solemn deed,
She scatters round the magic seed,
And thrice repeats: 'The seed I sow,
My true-love's scythe the crop shall mow.'
Straight, as her frame fresh horrors freeze,
Her true love with his scythe she sees.

And next, she seeks the Yew-tree shade,
Where he who died for love is laid;
There binds, upon the verdant sod
By many a moonlight fairy trod,
The Cowslip and the Lily-wreath

She wove her Hawthorn hedge beneath;
And whisp'ring, 'Ah! May Colin prove
As constant as thou was to love!'
Kisses, with pale lip full of dread,
The turf that hides his clay-cold head!

Author unknown, *The Cottage Girl*

Hemp may have something of a dubious reputation nowadays, but its story is rather a complex one. Historically grown for its fibre since the 6th Century, hemp was a crop once so common across the British Isles—particularly in fenland areas such as Cambridgeshire and Norfolk—that famous herbalist Nicholas Culpeper didn't even bother to describe it in his 1652 publication *The Complete Herbal*; it was impossible for him to perceive that someone could not know what it looked like.

Hemp fabric has been popular since the medieval period, enjoying a resurgence in the Victorian era and again today as people look for natural and renewable fibres. Nowadays, it's not uncommon to see hemp soap, facial products, oil, and even dietary supplements available to purchase. Until the late 1800s, ninety percent of all paper was made of hemp; even early drafts of the Declaration of Independence were likely to have been written on such. It was also a common sight in Victorian gardens, popularly recommended for the backs of garden borders as it provided a fast-growing, dense backdrop for smaller plants. The cultivation of hemp was so widespread, in fact, that escapees from historical fields can still be found growing wild, and in America it grows so abundantly that it is classed as an invasive pest and referred to as ditchweed.

A section from Thomas Tusser's A *Hundred Points of Good Husbandry* in 1557 highlights some of the common uses for this plant:

Wife, pluck fro thy seed hemp, the fimble hemp clean,
This looketh more yellow, the other more green;
Use t'one for thy spinning, leave Michell the t'other,
For shoe thread and halter, for rope and such other.

One of the most famous uses for hemp fibre was in the production of rope for execution. It became so associated with this purpose that the plant became a symbol of the gallows, and in Somerset in England was known as neckweed or gallow-grass. In the fenlands of Cambridge, if a man were to break the code of never betraying a fellow fenman, his door might be daubed by angry neighbours with an image of a stem of hemp and a willow stake, with the words: *Both grown for you*. The hemp he could use to hang himself, and the willow stake to be driven through his heart upon burial.*

The custom of driving a stake through a person's heart after death is particular to Cambridge and Norfolk, both fenland areas in the east of England. Staking is a famous way to deal with vampires, but in these areas it was also a punishment reserved for murderers and traitors. They would be buried in unhallowed ground at a crossroad, and the stake would stop their restless spirit from returning to trouble the living. Near to the Norfolk town of Harleston is an area known as Lush's Bush; there stood a willow tree there that was said to have grown from a stake driven through the heart of a local murderer named Lush. Though the tree was cut down in the 1800s, the area

* Enid Porter; *Cambridgeshire Customs and Folklore*

became the site for the burial of several criminals, and features heavily in local ghost tales.

Nowadays, *C. sativa* is best known for its use as a recreational drug. 'Hemp' is a term mostly reserved for commercial crops that have been grown specifically to contain only miniscule amounts of THC, the chemical that gives cannabis its intoxicating effects; the term 'cannabis' refers to those that are grown specifically for the recreational trade. Technically, however, they are still the same plant.

But this distinction is a modern one, and historically the plants grown for fabric and building materials were equally as intoxicating. The effects of THC did not go unnoticed, either. Cannabis has a long history of use dating back at least 5,000 years, and is mentioned in works by the Chinese Emperor Shen Neng in 2727BC. In 2019, traces of cannabis were found in tombs in Jirzankal cemetery in the Pamir Mountains, western China, which are at least 2,500 years old. The drug was found in wooden burners that would have been filled with leaves and hot stones to fill the area with smoke during the burial ritual.

This use is similar to the Scythian peoples, who, as recorded by Herodotus in his *Histories* in 430BC, would purify themselves after performing a burial by creating steam baths inside of a specially sealed tent. During these baths, they would throw hemp seeds onto hot stones to create a fragranced steam.

As with most plants that were common and essential to rural life, hemp became central to various folk tales and traditions

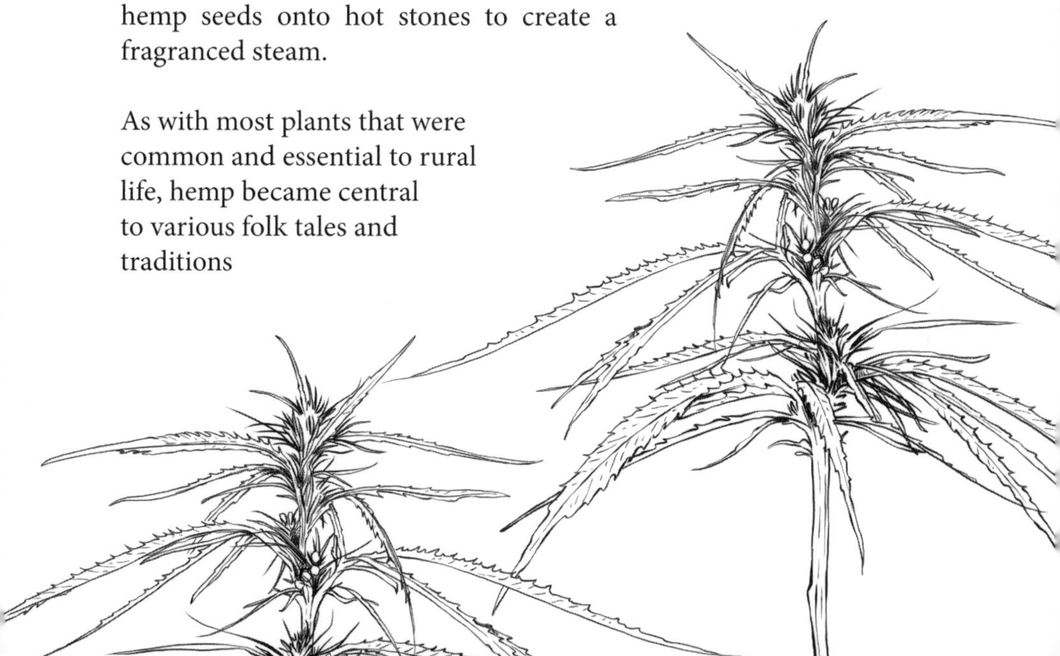

amongst those who farmed and used it. Many focused around the cultivation of the plant: in England, by custom, young women were not allowed to work in the hemp fields as it was thought that merely touching the plant would make them barren. In India, however, the Atharva Veda says that it is a protective herb—this 'plant with a thousand eyes' was created by Indra, and gifted with the ability to drive away disease and kill all monsters.

Another English belief concerning hemp gave rise to one of the best-known love divination rituals, which has been recorded regularly from 1685 onwards with no or little changes. It is also one of the more morbid of these tests to discover one's true love. The hemp seed ritual requires a girl to go out into a cemetery on Midsummer Eve, and throw behind her a handful of hemp seed without looking. Whilst doing so, she must recite:

Hemp seed I sow thee, hemp seed grow thee;
And he who will my true-love be, come after me and show thee.*

After reciting this, the maiden must run away from the spot. If she is brave enough to look, she will see the wraith of her lover pursuing her with a scythe. If she is fated to never marry, what pursues her will be an empty coffin, or a tolling bell. There is no record of what happens should she not manage to outrun these visions.

A legend in China relating to hemp speculates on the popular subject of what might befall humans when they encounter fairies. This story dates back to 60AD. It tells of two friends who, whilst wandering in the mountains, came across a fairy bridge. The bridge and its surrounding gardens were tended by two beautiful women, who invited them to cross the bridge and smoke hemp in the great city over the way. After several blissful days with their hostesses, the pair became homesick and decided to leave. However, when they did, they found that seven generations had passed and they had become old men. Unable to live in the mortal world in such old bodies, they turned into dust and disappeared.

* Charles Henry Poole; *The Customs, Superstitions, and Legends of the County of Somerset*

HENBANE, BLACK: Hyoscamus niger

Straightway venom wine-cups shout
Toasts to One whose eyes are out:
Flowers along the reeling floor
Drip henbane and the hellebore:
Beauty, of her tresses shorn,
Shrieks as nature's maniac:
Hideousness on hoof and horn
Tumbles, yapping, in her track.

George Meredith, *The Woods of Westermain*

This nightshade relative is an infamous one. Described in great depth in Greek sagas and western plays as a creation of great evil, henbane was once referred to as 'a poisonous and dangerous plant, of dismal aspect and disagreeable smell'.* And with its pale yellow-green flowers with purple veins, and the black centre that has given it the nickname Devil's eye, it looks as dubious as its reputation might suggest.

The name henbane sounds like a folk name, in keeping with plants such as dogbane and wolfsbane, but rather than referring to any poisonous tendencies towards chickens, it is thought that *hen* originally came from early roots of the word 'death'. The moniker dates back to at least 1265, which can make pinpointing the etymology difficult, but luckily this plant has many other names by which to identify it. An early Saxon term was *belene,* from *bhelena,* meaning 'crazy plant',† and in 8th Century Italy it was called *symphonica* after the instrument of the same name, which was a small rod, usually silver, hung with bells,‡ which the plant resembles.

Mysteriously, it is also attributed the name *deus caballinus,* or 'horse god', which supposedly dates back to the 13th Century. However, the only bonafide early reference to this is in a booklet

* Richard Brook; *New Cyclopaedia of Botany and Complete Book of Herbs, 1854*

† Henry Solomon Wellcome; *Anglo-Saxon Leechcraft: An Historical Sketch of Early English Medicine; Lecture Memoranda*

‡ Salvatore de Renzi; *Collectio Salernitana:* translated by George Corner in *The Rise of Medicine at Salerno in the Twelfth Century*

published by Pietro Castelli in 1638. In this year, Castelli founded a botanical garden adjacent to the city walls to serve the University of Messina. Eager to spread the word of the garden, Castelli published the *Hortus Messanensis,* a description of the garden detailing its contents and medical value. Unfortunately, the origin and reason for this curious name were never recorded.

The plant is also called also called hog's bean. Both the French name *Jusquiame* and the botanical name *Hyoscamus* are derived from the Greek *hyos* and *cyamus*, literally meaning 'hog's bean', as pigs are said to eat the plants without ill effect.

Though hogs may eat it with impunity, humans are unfortunately more vulnerable to the plant's toxins. All parts of the plant contain scopolamine and hyoscamine, which can cause hallucinations and restlessness. In larger doses, they can lead to convulsions, vomiting, delirium, respiratory paralysis, coma, and death.

In smaller doses, henbane was one of the plants commonly used by European witches to achieve the hallucinatory high that was used in 'flight' to the witch's sabbat. It was also reportedly used by the Oracle of Delphi to aid in prophetic visions. In 1955, German toxicologist Dr Will-Erich Penckert experimented with the fumes from henbane seeds in order to better report on their effects. Following is an extract from his notes:

I went to the mirror and was able to distinguish my face, but more dimly than normal. I had the feeling that my head had increased in size: it seemed to have grown broader, more solid ... The mirror itself seemed to be swaying, and I found it difficult to keep my face within its frame. The black discs of my pupils were immensely enlarged, as though the whole iris, which was normally blue, had become black. Despite the dilation of my pupils I could see no better than usual; quite the contrary, the outlines of objects were hazy.

There were animals which looked at me keenly with contorted grimaces and staring, terrified eyes; there were terrifying stones and clouds of mist, all sweeping along in the same direction. They carried me irresistibly with them. ... I was flung into a flaring drunkenness, a witches' cauldron of madness. Above my head water was flowing, dark and blood-red. The sky was filled with herds of animals. Fluid, formless creatures emerged from the darkness. I heard words, but they were all wrong and nonsensical, and yet they possessed for me some hidden meaning.

As well as prophecy and flight, henbane was notably remarked upon for being able to turn people mad. Bartholomaeus Anglicus wrote of it in his *De Proprietatibus Rerum* in 1240: 'This herb is called insana wood, for the use thereof is perilous; for if it be eate or dranke, it breedeth woodeness; therefore the herb is commonly called Morilindi, for it taketh away wytte and reason.' The use of *woodeness* is from the Old English word *wod*, meaning 'madness' or 'fury', from the name of *Woden*—more commonly referred to nowadays as Odin, a God known for his fits of fury. Another name for the plant was *alterculum*, used by the Romans, as those who partook of it would become angry and quarrelsome.

Despite its dangers in larger doses, henbane is not always necessarily deadly. In fact, early medicinal practitioners made use of it as an anaesthetic before its unpredictability saw it replaced by more reliable alternatives. Gerard noted this in his *Great Herball*: 'The leaves, the seeds and the juice, when taken internally cause an unquiet sleep, like unto the sleep of drunkenness, which continueth long and is deadly to the patient.'

More questionable characters also made use of this soporific property. In the 1300s, French travellers and pilgrims were in danger of having their drink or meal spiked with a mix of crushed henbane, darnel, poppy, and bryony seeds. Once they were asleep, they were at the mercy of thieves with an eye on their belongings.*

* John Arderne; *Treatises of Fistula in Ano*

Henbane's intoxicating properties also made it a popular addition to cheap beer. Like darnel, the juice of henbane would be added to watered down beer to produce the sensation of drunkenness, but at a fraction of the price to the establishment selling it. However, when the Bavarian Purity Law was passed in 1516, forbidding beer to contain anything other than hops, barley, and water (and later yeast), the practice ended.

Due to its poisonous nature, henbane has become intricately connected with funereal rites across Europe. Plutarch wrote of how Greek tombs would be decorated with chaplets of henbane, and the dead were said to wear these crowns as they followed the Styx into the Underworld. As they did so, the plant would make them forget their loved ones and the lives that they lead before death, so that they would not feel the desire to return. The tradition is not just a Greek one; traces of henbane have been found in Scottish Neolithic burial mounds, leading to speculation that the plant may have been used as an aid to guide spirits onwards.

Albertus Magnus, a bishop in the early 1200s, suspected that henbane gave necromancers power over the dead, claiming that burning it could invoke restless souls and demons. It also had the power to repel evil spirits: Italian midsummer festivals would see it burned to fumigate stables, to keep evil influence away from horses and cattle.

HYDRANGEA: Hydrangea spp.

These leaves are like the last green
in the paint pots—dried up, dull, and rough,
behind the flowered umbels whose blue
is not their own, only mirrored from far away.
In their mirror it is vague and tear-stained,
as if deep down they wished to lose it;
and as with blue writing paper
there is yellow in them, violet and gray.

Rainer Maria Rilke, *Blue Hydrangea*

Hydrangeas are a popular garden shrub, best known for their colour changing properties: depending on the acidity of the soil they are grown in, the large flowering heads can bloom in blue or pink, or even (with some clever gardening) be encouraged to display both colours at once. Acidic soil that's heavy with aluminium produces the blue flowers, and lime-heavy soils produce pink. The 'magic' of the changing flowers is said to be a gift from the fair folk, who change the colours depending on their whims; blue is said to be the luckiest, with pink a warning of bad fortune to come.

Despite its popularity, the plant actually contains low levels of cyanide, which can, in large enough doses, cause nausea, vomiting, and sweating. Fortunately, it does not contain enough to pose a threat to most gardeners.

In the Victorian language of flowers, it represents 'heartlessness'. Planting hydrangeas at your door was said to doom your daughters never to marry, and Victorian men would send bouquets of hydrangeas to women who had turned them down, perhaps in the hope of bestowing the same curse on them.

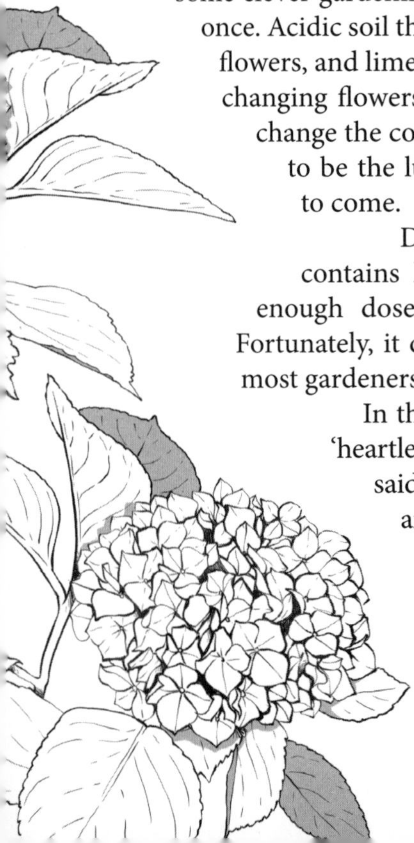

IVY: Hedera helix

Ivy is soft and meek of speech,
Against all bale she is bliss;
Well is he that may here reach.
Veni, coronaberis,

Ivy bears berries black,
God grant us all His bliss.
For there shall we nothing lack,
Veni, coronaberis.

Medieval Carol, 1430, *A Song of Great Sweetness from Christ to his Daintiest Dam*

Ivy is, perhaps, the quintessential plant of Europe. A skilled and determined climber, ivy can establish itself on any rough surface that will help it reach a sunny spot, and in some areas it's hard to find a wall or tree or fence that isn't heavy with the weight of this vine. Although evocative of Christmas, of cemeteries and ancient woodlands, ivy is damaging to the hosts that it grows on, smothering trees and breaking into bricks and stones with the small roots that it uses to climb. Ironically, in the case of old buildings, it can be the only thing still holding the brickwork together. And yet, it is one of the most essential plants to upholding a healthy environment for small wildlife; it provides protection for birds and endangered species of bats, and food for thousands of insect and moth species.

Though native to northern Europe and some areas of southern Asia, *H. helix* was introduced to North America and Australia by English colonists, and is still known as English Ivy in many countries. In England itself, it is also known as bindwood and lovestone, for the way that it voraciously climbs and engulfs anything in its way.

Ivy's main goal is to soak up as much sunlight as possible, so it is prevalent in open places with plenty of warm surfaces to support it. As a result, one of the places it grows best is in graveyards and across old tombstones, which has created a connection to death and mortality that persists even today. An old English belief was that an absence of ivy on a grave meant that the soul was a restless one; but if a grave belonged to a young woman and was bequeathed with an abundance of the plant, it was a sure sign that her death was from a broken heart.

This connection between love, death, and ivy is illustrated best in the medieval legend of Tristan and Isolde, two lovers who were unable to be together whilst they lived. After their deaths they were buried separately; but ivy grew between the two graves, ensuring that they would always remain connected. No matter how people tried to cut or remove the vines, they would always regrow to be together.

As with most plants associated with death, superstitions prevail that the ivy vine is unlucky to bring into a house. This is a belief that, once again, began in England,* and has travelled even as far as records in Alabama and Massachusetts in the US.† Nearby in Maine, the act of bringing ivy into a home is supposedly a sign that the homeowner will always be poor.‡ There is, however, one time of year when it is fortuitous to bring ivy into the house: Christmas, so long as it is gone again before Candlemas day (February 2nd). Since the land outside of the house is cold and inhospitable even to the spirits and fair folk who dwell there, bringing ivy inside is an invitation for those displaced creatures to take shelter in the warm. Since they have no power to cause harm during the holy month of December (no doubt an addition made by the Christian

* Enid Porter; *Some folk beliefs of the Fens. Folklore. Vol 69*

† Ray Browne; *Popular Beliefs and Practices from Alabama*

‡ Fanny Bergen; *Current superstitions collected from the oral tradition of English speaking folk. American Folklore Society. Memoirs. Vol 4*

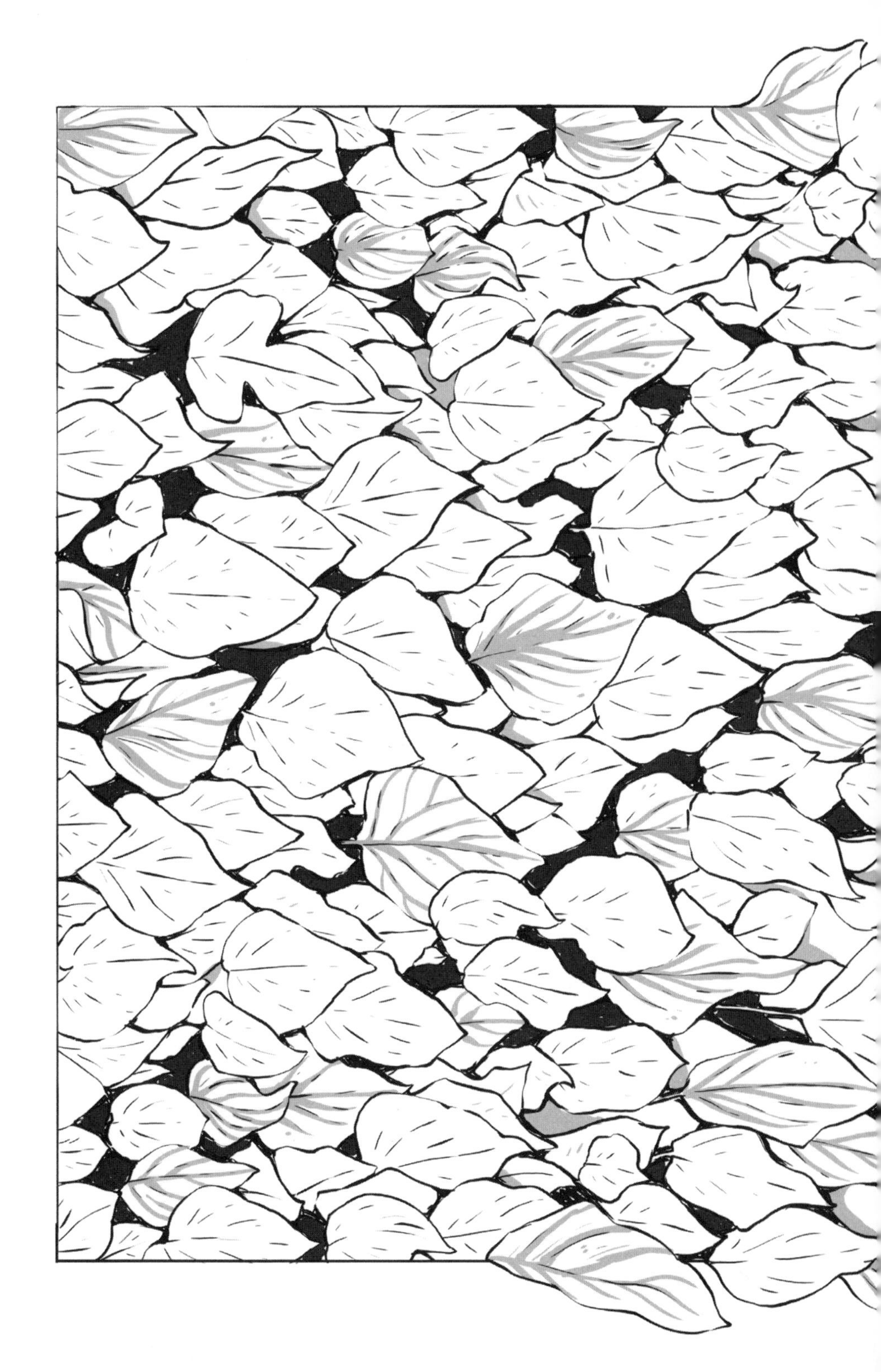

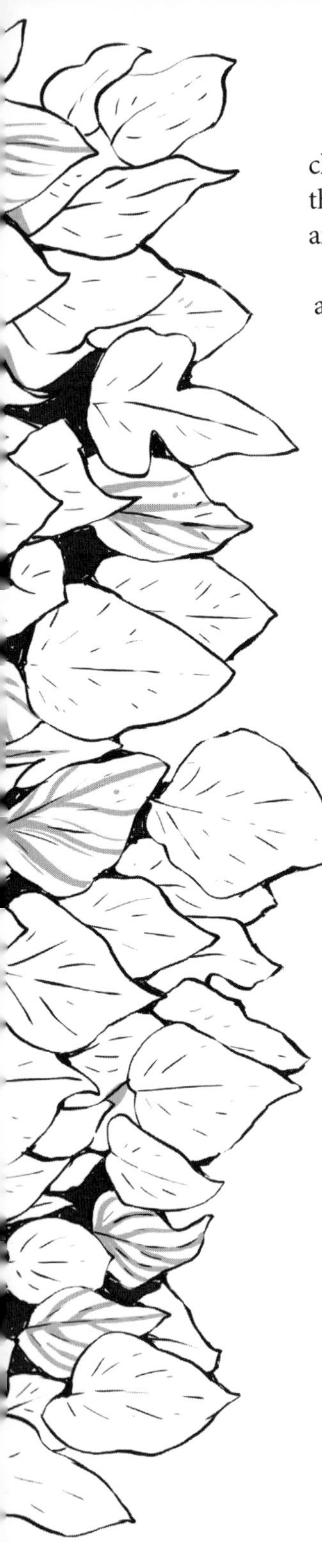

church when much early folklore was absorbed into the liturgical calendar), it is a safe time to offer shelter and win their favour.

This long-prevalent superstition is reflected also in Finland, where remaining in the favour of forest spirits is a key message in many old folk tales. The concept of forest enchantment features in myths across the globe, and in Finland it is called *metsänpeitto,* the forest blanket. It is used to describe the unexplained disappearance of people or domestic animals.

Forests cover approximately seventy-eight percent of Finland, and many old Finnish expressions compare the forest to a church; it is a sacred place, not to be underestimated or mistreated. *Niin metsä vastaa, kuin sinne huudetaan:* 'The forest answers in the same way one shouts at it'. Those who do mistreat the forest, or wander where they are not supposed to, are at risk of becoming 'covered by forest'. Similar to the way that fairies in the British Isles might steal away those who step into a mushroom ring or a fairy hill, the forest is full of mischievous creatures known as *maahinens,* small gnomes who spirit away the unwary. Records from those who claim to have been prey to the *metsänpeitto* describe not being able to recognise areas that they know should have been familiar to them, or being invisible to others around them, or being unable to move or speak. In most of the records we have of this phenomenon occurring, the forest is said to be deathly silent.

The descriptions and concept of *metsänpeitto* is similar to the Japanese concept of *kamikakushi* (literally 'hidden by spirits'), or being 'spirited away'.

In both concepts, a method of escape is to do something which confuses the spirits, such as pouring water into your footprints, wearing clothes inside out, switching shoes to the wrong feet, or carrying plants with protective properties, such as ivy.

Ivy might provide protection to those lost in the forests, but it is still, at its heart, a poisonous plant. The leaves and berries of English ivy contain a-hederin, a glycoside that can cause vomiting, convulsions, and muscular weakness.

The leaves also produce a narcotic effect not unlike that of atropine, which, despite their toxicity, have made them an unlikely favourite for including in alcoholic drinks. In Oxford, UK, an interesting tradition amongst the college students still exists. Every year at 11am on Ascension Day (39 days after Easter Sunday), a small tunnel that links two of the colleges, Brasenose and its neighbouring Lincoln College, is unlocked, allowing Brasenose students to enter the Lincoln bar for a free pint of ale that has been spiked with ivy leaves. The origin of this tradition is a typical urban legend; it comes with no names or dates to corroborate it, with several variations on the narrative besides, but each person who recounts it to you will be absolutely certain that their version is the truth. The best lead we have is that the 'ivy ale' tradition appears in records as far back as the 1700s.

Supposedly, at some point in time a Brasenose student was being chased by an angry mob. Some versions claim that the mob was a group of local townsmen—whom Oxford students are infamous for antagonising—whereas others say that the pursuers were students from Balliol, a rival college. Whomever the unfortunate student was fleeing from, he found his way to Lincoln College, and begged to be allowed in. When the college refused, he was captured and murdered by the mob. In subsequent years, as an apology for having allowed this to happen, Lincoln allowed Brasenose students to enter once a year to have free access to their ale supplies. However, the cost of this soon began to add up (as well as the inconvenience of having their supplies drunk dry each year), and so Lincoln began adding crushed ivy leaves to the ale that they served, hoping that the bitter taste and stomach aches would deter the incoming Brasenose students from drinking too much. Presumably it did not work as intended, as the tradition still continues today.

But the inclusion of ivy in alcohol goes back further than college students and an angry mob: it dates back as far as the Greeks… and an angry mob. The ivy vine was sacred to Dionysus, the Greek god of wine and agriculture. Female followers of Dionysus, or Bacchus as he was also known, were called *maenads* ('raving ones') who would celebrate their god by drinking wine and chewing ivy leaves to drive themselves into a state of frenzy. Wearing snakeskins and fox pelts, they would rampage across the countryside, attacking animals and humans alike and pulling them apart with their bare hands. It was a particularly bloodthirsty sect of the cult that sparked a crackdown on such revels, starting a hunt in 184BC that led to almost two thousand women being tried and executed on charge of poisoning.

Bacchus himself was said to have been abandoned as a babe underneath an ivy bush, which was subsequently named after him. He wore a crown of corymbi, ivy berries, and his followers would wear wreaths of ivy leaves and tattoo themselves with the same design. *The ivy wreath became associated with alcohol and drinking, and any pub that offered ale or wine would hang a pole wreathed in an ivy bush outside its doors to advertise that alcohol could be bought within. An old idiom—*good wine needs no bush*—comes from this practice, suggesting that any place with a reputation for good drinks does not need to advertise it.

* Walter Otto; *Dionysus: Myth and Cult*

LILY OF THE VALLEY: Convallaria majalis

Of flowers whose fragrance freights the vernal gale,
Whose colour blends on Flora's banners fair,
What scent, what form, sweet lily of the vale,
Can in the eye of taste with thee compare?
But I will bear thee from that lowly bed,
Where 'mid involving shades thy blossoms rise,
Gem of the morn! And round Aurelia's head
Shall twine thy wreath in grateful sacrifice.

G G Richardson, *A Basket of Lilies*

Lily of the valley is a highly popular garden plant, prized by many for its sweet scent and tiny, white, bell-shaped flowers that are some of the first to appear in spring. In the Victorian language of flowers, it was given to represent peace, happiness, and harmony, and was dedicated to the Virgin Mary by the Christian church; the French call it *Larmes de Sainte Marie*, 'Our Lady's Tears'. It is said to grow best when planted near solomon's seal, a plant supposed to be the husband of the lily of the valley. This belief likely arose due to the lily's mention in Solomon's *Song of Songs*, a section of the Old Testament:

She:
I am a rose of Sharon,
a lily of the valleys.

He:
Like a lily among thorns
is my darling among the young women.

Although beautiful and innocent in appearance, all parts of this plant are highly poisonous, particularly the red berries that grow after the flowering season. Even in small amounts, ingestion can cause vomiting, reduced heart rate, blurred vision, and abdominal pain.

Despite its toxicity, it has continued to appear in folk medicine and historical texts alongside mentions of its supposed medicinal value. In the way of the doctrine of signatures—an early pseudoscience which claimed that plants could heal the parts of the body they resembled—the heart-shaped seeds of the lily of the valley were used in Russia to treat heart conditions, despite no evidence to confirm their effectiveness. John Gerard, a herbalist whose *Great Herball* became the go-to encyclopaedia of plants and their uses in the 1600s, even claimed that the plant 'without doubt, strengthens the brain and renovates a weak memory'. However, as well as being famous for the *Herball* he was also well-known for fabricating a great deal of its contents, and there is no proof that his writings were actually of much value in this regard.

The only way in which the flower may have proven valuable in the medicinal field was against stroke and nervous disorders. *Agua Aurea,* or 'gold water', was distilled from the plant, and was supposed to be so effective against strokes (known at the time as apoplexy) and so valued that it was kept in vessels of gold and silver. It was also used as an antidote to poison gas in the First and Second World Wars due to its ability to slow the heart. Nowadays there are many safer alternatives available and the plant has no modern medicinal value.

The plant's delicate appearance, mirrored against its dual nature as a poisoner, has made it a firm favourite in folk tales across Europe where it originated. Along with other white-flowered plants such as snowdrops and white lilac, it is believed to be a harbinger of death or bad fortune. In Devonshire in the UK it is believed to bring death within a year to anyone who plants a bed of these flowers, whereas in the county of Somerset, if brought into the house it will bring death with it, particularly for any young girls who live there. A local folk tale known as *The Basket of Lilies* tells of a woman who loved the plant so much that she would send her daughter out to pick them and bring them home. Everyone warned her of the danger, but she did not heed it, and her daughter eventually sickened and died.

Two other stories tell of how the lily of the valley came to be, and why it blooms at the time of year that it does. In Sussex in England, legend tells of a series of battles between Saint Leonard and a dragon that plagued the area. Each battle drove the dragon further and further back into the forest, where at last it disappeared. Every year, the locations of the battles are revealed by great numbers of lily of the valley, which grow where Saint Leonard's blood fell.

The other legend is somewhat more bittersweet. It tells of how the lily of the valley used to bloom all year round, until it fell in love with a nightingale that filled the woods day after day with its beautiful song. But no matter how much the lily pined, she was too shy to profess her love, and when winter came the nightingale left the woods and the lily behind. Broken-hearted, the lily stopped blooming, only showing her flowers again once the nightingale returns each May.

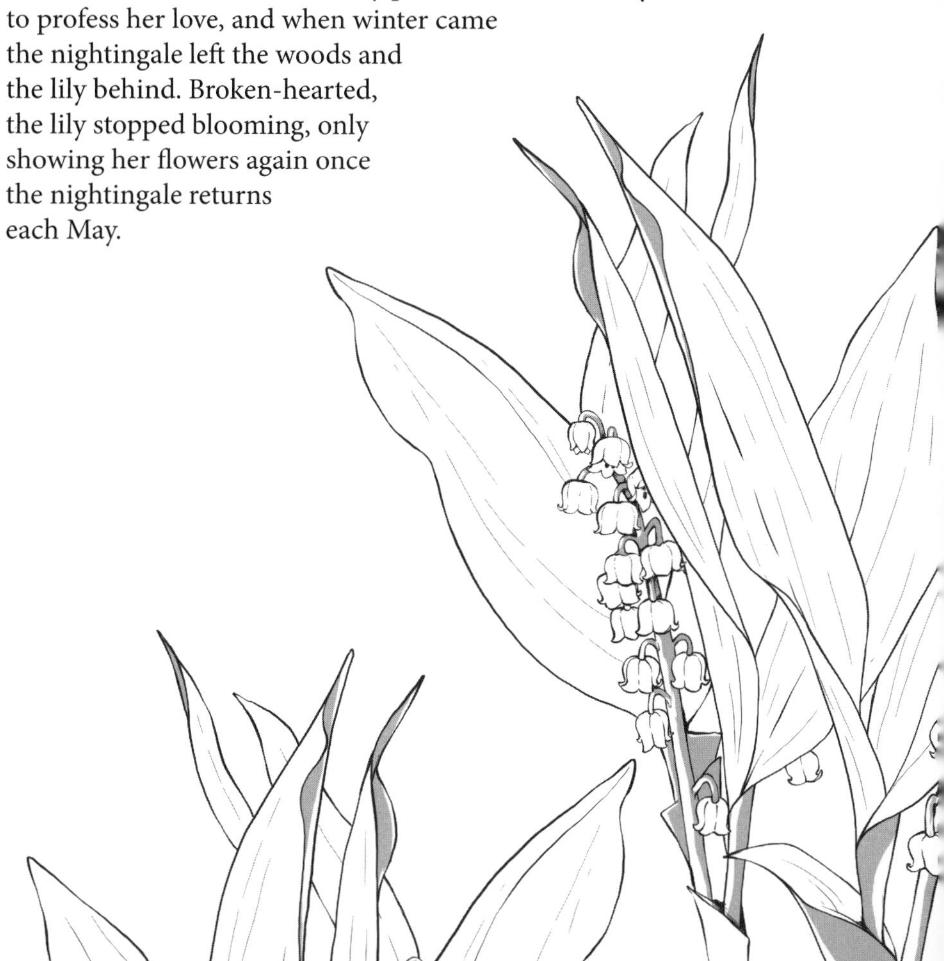

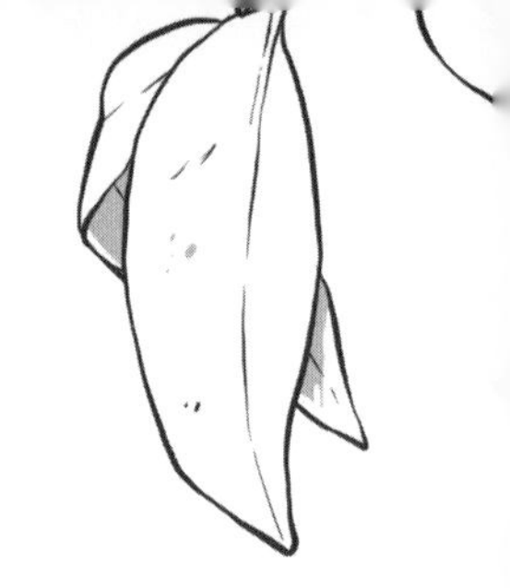

MANCHINEEL: Hippomane mancinella

Your gentle perfume, they say, gives fatal bliss
which for a moment transports one to Heaven
and then brings on the slumber without end.

Giacomo Meyerbeer, *L'Africaine*

Also known as the poison guava, manchineel is a tree native to Florida that in 2011 took the Guinness World Retcord title for being the most dangerous tree in the world. Most of the danger comes from its innocuous appearance; to the untrained eye it is just another fruit tree, laden with sweet green apples.

But these beach apples, despite their 'pleasantly sweet' juice and similar taste to plums (according to radiologist Nicola Strickland, who wrote of her experience after eating one), can cause the throat to swell closed, impeding breathing and leading to death. Johann Zahn described them as such:

*There is a tree in Hispaniola, bearing Apples of a very fragrant smell which, if they are tasted, prove hurtful and deadly. If anyone abides for a time beneath its shade he loses sight and reason, and cannot be cured save by a long sleep.**

* Johann Zahn; *Speculae Physico-Mathematico-Historica Notabilium ac Mirabilium Sciendorum*

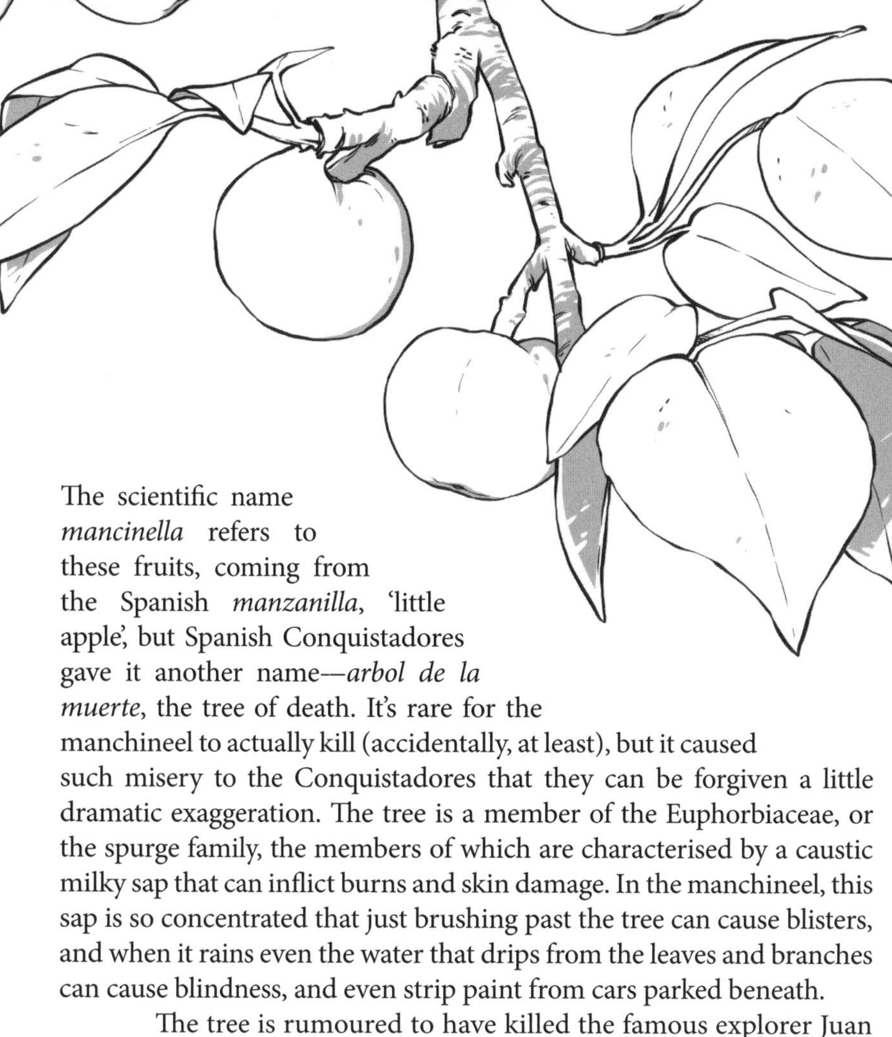

The scientific name *mancinella* refers to these fruits, coming from the Spanish *manzanilla*, 'little apple', but Spanish Conquistadores gave it another name—*arbol de la muerte*, the tree of death. It's rare for the manchineel to actually kill (accidentally, at least), but it caused such misery to the Conquistadores that they can be forgiven a little dramatic exaggeration. The tree is a member of the Euphorbiaceae, or the spurge family, the members of which are characterised by a caustic milky sap that can inflict burns and skin damage. In the manchineel, this sap is so concentrated that just brushing past the tree can cause blisters, and when it rains even the water that drips from the leaves and branches can cause blindness, and even strip paint from cars parked beneath.

The tree is rumoured to have killed the famous explorer Juan Ponce de Leon on his second trip to Florida in 1521. At the mercy of the Calusa people, the native tribe that lived in Florida's South West, de Leon and his men suffered many attacks from the natives, who dipped their arrowheads in the latex of the manchineel, tied captured enemies to its trunk, and used the leaves and bark to poison their wells. The Spanish became so frustrated with the dark powers of this tree that in their reports of the expedition they claimed that a person who even sat or walked beneath it would end up blinded or dead.*

* Paul Standley and Julian Steyermark; *Flora of Guatemala*

MANDRAKE: Mandragora officinarum and M. autumnalis

The phantom shapes—oh touch them not—
That appal the maiden's sight,
Lurk in the fleshy Mandrake's stem
That shrieks when plucked at night.

Thomas Moore, *Untitled*

Of all of the usual suspects when discussing plants and their magical, poisonous, or historical attributes, mandrake is among the most likely to be mentioned. Famous for its forking, humanoid roots and otherworldly scream, and long-respected as a powerful poison and tranquiliser, mandrake's endless list of uncanny tales and magical abilities has captured imaginations for centuries.

A member of the Solanaceae family alongside henbanes, nightshades, and daturas, the earliest scientific name of this plant was *Atropa mandragora,* after the eldest of the Greek Fates, Atropos. The Greeks knew it as Circeium, after Circe, the goddess of witchcraft and poisonous herbs. Nowadays, the plant is known as *Mandragora officinalis. Mandragora* is named for the mischievous, dark-skinned creatures of the same name (meaning 'man dragons') that were thought to possess the mandrake. Many plants with medicinal properties have the specific epithet *officinalis,* the Medieval Latin used to denote medicinal herbs. It literally means 'belonging to an officina', a monastery storeroom where medicines would be kept.

In countries where mandrake does not natively grow, there has historically been some confusion with other plants that have some similarity, usually in the size or shape of the roots. In the British Isles, plants such as bryony (both black and white), cuckoo pint, and enchanter's nightshade have been locally called by the name of mandrake, and become mixed up in many of the tales associated with the true specimen.

All species of actual mandrake contain tropane alkaloids that can cause hallucinogenic effects, as well as damaging the nervous system and causing dizziness, vomiting, and a rapid heart rate. The roots, on top of these effects, have narcotic properties, and it is for these that the mandrake was used as an early anaesthetic device known as the dwale. A dried sponge would be soaked with the juices of narcotic plants such as henbane, hemlock, and mandrake, and later rehydrated in hot water for a patient to inhale in order to render them unconscious.

Beyond the medical field, mandrake's soporific properties have been used more than once for the business of war. Among tales that circulated in the Roman high command is one regarding the great general Hannibal (or in some versions, Maharbal, one of his officers), who made use of the root whilst subduing an African rebellion near Carthage. Knowing that the rebels would invade their encampment if they pretended to retreat, Hannibal drugged his own army's supply of wine with mandrake root and then pretended to abandon the base as if in a hurry. The rebels came in and, believing themselves victorious, drank the wine to celebrate—allowing Hannibal and his force to return to slaughter or capture the drugged and sleeping men.* A second tale tells of a remarkably similar plot involving Caesar and a group of Cilician pirates who had taken him captive.

There are also records of mandrake root being put to personal use as a sleep aid. A 12th Century manuscript discusses mixing the powdered bark of the root with egg whites and applying it to the forehead to improve sleep,† a method that may have survived until Shakespeare's time, as in *Anthony and Cleopatra,* the Queen asks her maid for 'mandragora' to help her sleep during Anthony's absence.

Despite its historical uses, the tales about the mandrake that endure best are the more fantastical ones. An undeniably strange plant, it has been the subject of many contemporary writers, and long before them tantalised the superstitions of people in every country where it grows. Numerous are the books and herbals that will tell you it is a plant of the Devil, growing rampant beneath the gallows and crossroads where murderers, suicides, and witches were hanged and buried. Some even claim that the roots lead all the way down

* Sextus Julius Frontinus; *The Stratagems*
† Pietro de Crescenzi; *Ruralia Commoda*

to Hades' underworld, and that if you're not careful, you might fall down there after pulling up the plant.

Most famous, of course, are the tales that pertain to the supposed humanoid nature of the root itself. They are naturally inclined to splitting and forking, particularly in the stony soil the mandrake favours, and when the plant is uprooted can (with a degree of imagination) appear to be small, warped humanoids. In Dioscurides' early writings, he talks at length about the male mandrake and female mandrake; we now know that he was speaking of two different species, *M. officinalis* (male) and *M. autumnalis* (female), but these early sources of misinformation have only served to add fuel to the fire.

Across Europe and the Middle East are tales of these little human forms and the power they contain. Talismans made of part or the whole of the root would be wrapped in sheets of cotton and carried to bring good luck, with the understanding that whomever purchased it was entering a contract, and that the spirit of the root would be bound to the owner until their death. As a result, the talisman could never be given away; if it was, it

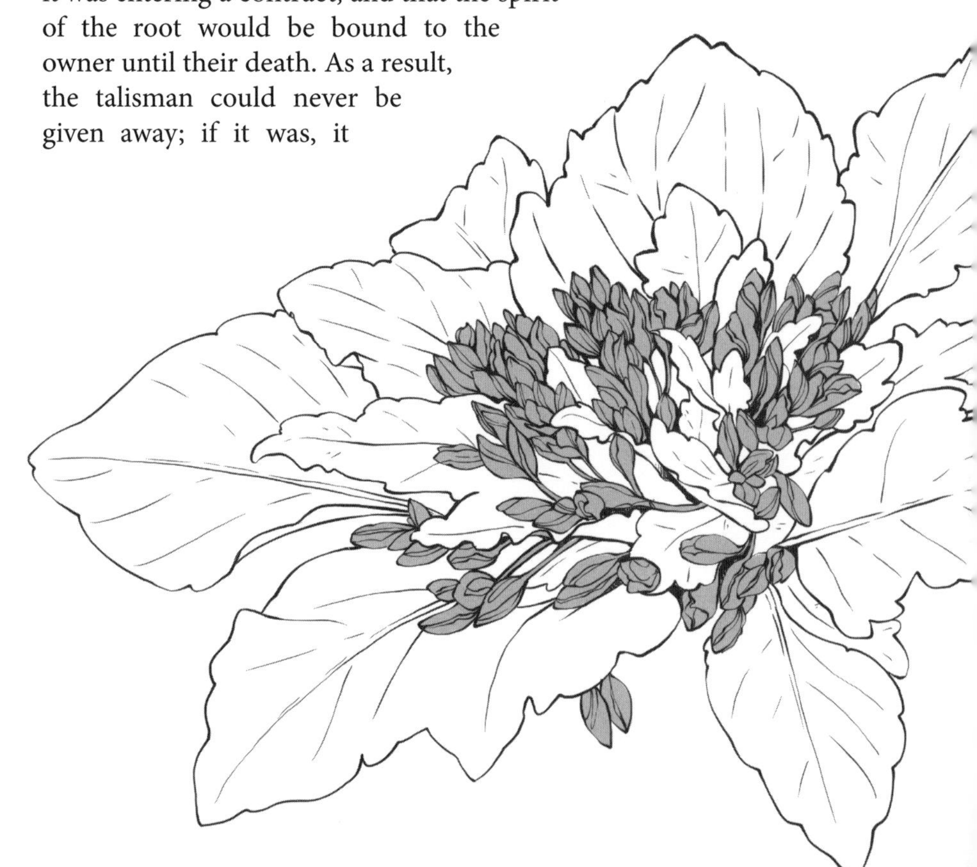

could only be sold and not gifted, and for less than what it had been purchased for originally.*

Others believed that it wasn't just the humanoid shape of the root that made it special: the root was surely a little person in and of itself, usually referred to as *mandragora*. Germans use the word *alruna* to mean both 'witch' and 'mandrake', and believed that witches could make from the roots a creature called the *alraun.*† This devilish spirit could reveal secrets, remove enemies, and double any coin given to it, but with one warning: if you overworked it, it would die.‡ Similar was a French belief that mandrake grew at the foot of oaks wreathed in mistletoe, the roots as deep in the earth as the mistletoe was high in the tree. The man who discovered it was obliged to feed it meat or bread every day, and if he ceased in this task, the *mandragora* would kill him. However, his services would not go unappreciated, and whatever was given would be returned twofold the next day.

An interesting variant on the *mandragora* is one recorded by Stéphanie Félicité, an 18th Century writer better known by the moniker Madame de Genlis. She speaks of *mandragora* as a sprite that would hatch from an egg that had been incubated in a particular manner, appearing as a little monster that seemed half-chick and half-man. This creature had to be kept secret and fed with seeds of the spikenard plant, and in return, every day it would provide a prophecy for the future.

Another persisting myth about the root is the terrible shriek it is said to emit when unearthed. The scream was said to be so piercing that it would immediately kill the one who unearthed it, and this is a tale that has been repeated through historical herbals, plays, and modern literature alike. For those nonetheless determined to unearth the plant, a common trick to avoid certain death was to tie a rope about it, then to secure the other end of the line to a dog. The hope was, since it was the dog and not the forager who had done the deed, death would fall upon the animal instead.

* Charles Skinner; *Myths and Legends of Flowers, Trees, Fruits and Plants*

† H. F. Clark; *The Mandrake Fiend, Folklore Vol. 73*

‡ James Frazer; *Jacob and the Mandrakes*

The truth is that the root *does* emit a small squeak when pulled from the ground, as most tuberous plants do, but there are no real reports, of course, of anyone who has been killed by the scream of a mandrake plant. Most likely the stories of its fatal shriek were spread by those who had genuine need of the plant; mandrakes need at least two or three years to fully mature to be of medicinal use, and no doubt a horror story of certain death to unsanctioned foragers was enough to protect at least a few of the juvenile plants from eager scavengers.

An equally farfetched tale—but one with a possible basis in truth this time—is the belief that the plant shines at night, a superstition that is so pervasive that the mandrake has been referred to as 'Devil's candle' in both Arabic and English manuscripts. Thomas Moore described it thus in *Lalla Rookh*:

Such rank and deadly lustre dwells
As in those hellish fires that light
The Mandrake's charnel leaves at night.

The earliest reference to this belief is a record from the 1st Century historian Flavius Josephus, who described a plant that grew in the castle of Machaerus, Jordan. The leaves were said to be the colour of flame, and to shine like lightning until it was approached, at which point the illumination would vanish. Less than a hundred years later, the Roman author Aelian described a magical, brightly-coloured herb with similar properties known as *Aglaophotis*, as it shone like a star at night. These are not the only two known references to a plant that shines in the darkness, and it is believed that these all refer to the mandrake. Not only do many of the attributes given to this magical plant line up accordingly, but mandrake leaves are also particularly attractive to glow worms, which would well explain the mystical vanishing lights.

MAPLE TREE: Acer spp.

And when her leaves, all crimson,
Droop silently and fall,
Like drops of life-blood welling
From a warrior brave and tall,
They would tell how fast and freely
Would her children's blood be shed
'ere the soil of our faith and freedom
Should echo a foeman's tread.

Henry Faulkner Darnell, *The Maple*

Famous for taking pride of place on the Canadian flag and for the ninety million kilograms of syrup produced yearly, the maple tree is recognised across the globe. But for all that it is largely synonymous with Canada's most famous export, maple trees are mostly native to Asia, and are found across Europe and northern Africa as well as North America. Almost all maples turn the vibrant red that the species is known for, and in Japan and Korea, events called *momijigari* and *danpung-nori* respectively are fondly-observed occasions on which to view the changing colours.

The name *Acer* means 'sharp', referring to the neat points of the star-shaped leaves, but it is also appropriate for a historical use of the wood: it is hard and easily shaped, which made it an ideal wood for spears and lances amongst many North American tribes. The Algonquian tribes particularly favoured this tree, and it was through them that Canadian settlers learned how to make maple sugar and maple syrup, the art of which the Algonquians had perfected for centuries. Maple sap was considered a gift from the Creator or from hero figures in various mythologies across the area, and many aspects of Algonquian tradition revolve around maple trees and the art of collecting the sugar.

In ancient Greece, the vibrant red of the maple leaves set it apart from other trees. It was believed to be under the powers of Phobos and Deimos, the twin personifications of fear and terror,* who would ride into battle alongside Enyo and Eris, the goddesses of war and discord. Phobos and Deimos's worshippers made many bloody sacrifices in their names, as they were said to delight in blood, and so thirsty were they for death that they were said to have a temple built from the skulls of those who had been slain in their name.

Another legend about the maple tree tells of how a young woman was transformed into a maple after death. This story comes from the historical region of Moldavia, now Romania and Ukraine, where the native species is *A. rubrum*, the red maple.

The story tells of a lord's youngest daughter, who fell in love with a young shepherd after hearing him play the flute. When spring came, the lord sent his three daughters to pick strawberries, and promised that whomever returned first with a basketful of fruit should inherit his lands. The youngest daughter was the first to complete the task, and her sisters, unwilling to rescind their birthright, murdered her and buried her body beneath a maple tree.

The two elder daughters returned home, and told their father that their sister had been killed by an elk. The lord grieved, as did the shepherd, and no matter how he tried to play his beloved flute it would make no sound. On the third day of grieving, the shepherd noticed that a new sapling had begun to grow at the base of the maple tree in the fields. Cutting it down, he fashioned from it a new flute, but the moment he put

* Diana Wells; *Lives of the Trees: An Uncommon History*

his lips to it, it began to sing instead. 'Play, beloved! I was a lord's daughter, then I was a maple shoot; now I am but a wooden flute.'

Astonished by this revelation, the shepherd rushed to the lord and told him what had happened. The lord put the flute to his own mouth, and it began to sing: 'Play, father! I was a lord's daughter, then I was a maple shoot; now I am but a wooden flute.'

The lord was convinced that he must have misheard the flute, and so he called in his two remaining daughters and demanded that they also try it. As each did so, the flute sang: 'Play, murderer! I was a lord's daughter, then I was a maple shoot; now I am but a wooden flute.'

Realising what must have happened, the lord exiled his daughters for the rest of their days to a barren island in the Black Sea. The shepherd returned to his fields, able to hear his love's voice only when he played the flute.

MISTLETOE: Viscum album

If she had been the Mistletoe
And I had been the Rose—
How gay upon your table
My velvet life to close—
Since I am of the Druid,
And she is of the dew—
I'll deck Tradition's buttonhole—
And send the Rose to you.

Emily Dickinson, *If She Had Been the Mistletoe*

European mistletoe is intrinsically entangled with stories of spellcraft, Druidry, and of course, the season of Yule and Christmas. The connection with mistletoe and winter holidays hails back solely to druidic influences: as with all evergreens, mistletoe is connected with the idea of immortality, and life in the midst of death, as any plant that can grow when others cannot must be able to defy death itself. In the historic Germanic region of Holstein, it was called 'spectre's wand', as druids were said to be able to see and speak with ghosts whilst holding a branch of mistletoe.

Over the years stories of druids have been exaggerated or noted without context, and though Druidism still exists as a spiritual movement the concept of a 'druid' in folklore has often come to simply mean one who has magical abilities, or follows traditions of older and less regulated faiths. One embellished story of the inexplicable power of druids is set in England, where mistletoe grows prolifically in the south and west of the country—except in the county of Devon which, according to this tale, has been cursed by druids so that it will never grow there. A particular orchard is said to straddle the Devon and Somerset border, where the apple trees on the Somerset side are laden with mistletoe, but the Devon side is completely devoid of it.*

Wherever stories such as these have originated from, mistletoe has always been considered magical. This parasitic plant is, by nature, a

* *Milleducia: A Thousand Pleasant Things Selected From Notes And Queries*

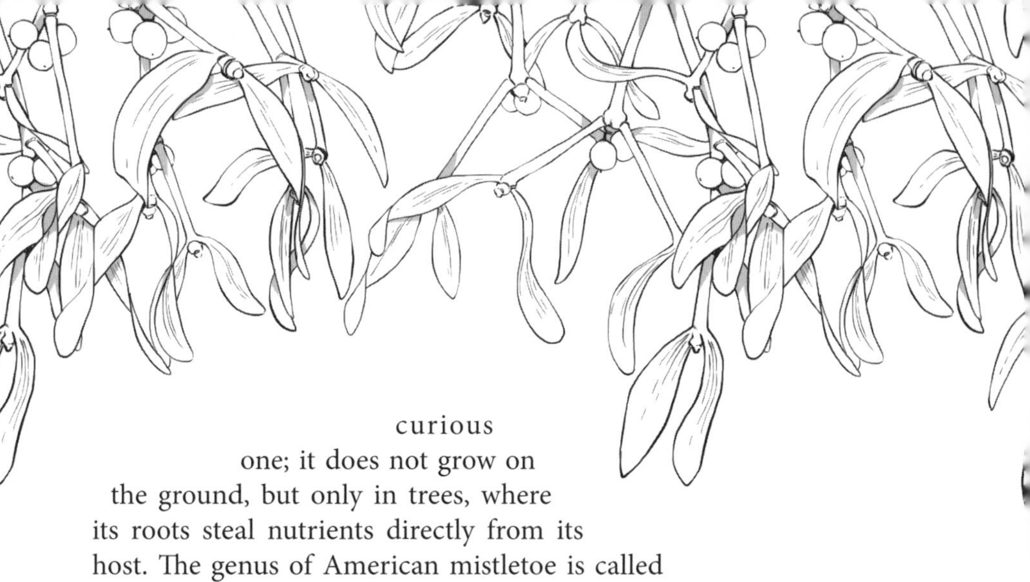

curious one; it does not grow on the ground, but only in trees, where its roots steal nutrients directly from its host. The genus of American mistletoe is called *Phoradendron*, literally 'tree thief', and for many centuries it was believed that mistletoe must have sprung from branches where birds had left their droppings. Its name still provides a nod back to that belief—*mistel* is the Anglo-Saxon for 'dung', and *tan* means 'twig': therefore, the modern day common name for this plant translates literally to 'dung on a twig'.

Birds may have been credited with propagating the plant, but mistletoe was once also used to cause great damage to the bird population. The juice of the mistletoe berry is clammy and viscous, and was a core ingredient in creating birdlime. This substance was used up until the 16th Century to capture small birds, as branches coated in it were sticky enough to entrap their feet. Some variations on the mixture were even strong enough to capture hawks, which could be lured onto a branch by tethering a small live bird to it.

Though the origin of the scientific name *viscum* is widely speculated over, it has been suggested that it may refer to this adhesive, viscous nature. Another theory is that it may originate from the Sanskrit *visam,* meaning 'poison'. The berries are toxic, but not hugely so; ingestion of the berries has never been recorded as causing long-lasting effects, but can cause symptoms of drunkenness that pass within a day or two.

The most famous tradition involving this plant is the act of kissing beneath it in the winter season. Nowadays, catching people unawares beneath it is mostly seen as a harmless party prank, but historically it was a much more serious business; if a couple kissed beneath it, it was intended as a promise to marry in the coming year. Traditionally, with each kiss that was shared beneath it, a berry should be plucked from the branch, and once all berries were gone there would be no more kissing. If any person remained after this who had not been kissed, they should remain single for the coming year.

Its nature in Scandinavian countries is a little less romantic. Associated with war, its forked branches resemble lightning and as a result it is closely linked with stories of the Norse god Thor. In Sweden it is known as 'thunder-besom', a name that remains from an earlier tradition where those who suffered epilepsy might carry a sprig of mistletoe, or a knife with a handle of mistletoe wood, to prevent them from being struck down by the inner 'electrical storm' of an epileptic attack.

A Norse legend tells of how mistletoe was responsible for the death of Baldur, the god of light and son of Odin and Frigg. Said to be beautiful, just, and beloved of all the gods, Baldur became troubled by dreams that he would die. To ensure that this would never happen, Frigg visited every living thing on the earth, and made them swear that they would do no harm to her son. By mistake she overlooked the mistletoe, and when the trickster god Loki learned of this, he fashioned a plan to be rid of Baldur.

Baldur, having realised that he was now invincible, had made a great game with the other gods, inviting them to hit him with their weapons to prove that he could not die. Loki, fashioning an arrow from mistletoe wood, encouraged the blind god Höd to fire it at Baldur, convincing him that it would do no harm. But the arrow slew Baldur, and his dreams came to fruition. The pearly white berries of the mistletoe are said to be the tears that Frigg shed upon realising that her precious son was dead.

OLEANDER: Nerium oleander

> Like a Pharisee, that is beauteous without,
> and within, a ravenous wolf and murderer.
>
> William Turner, *A New Herball: Parts II and III*

The oleander is an ornamental flowering shrub that thrives in tropical climates. It is so popular as a cultivated plant that it is now hard to pinpoint its exact country of origin, but it is thought that similar wild varieties may have come from southwest Asia. The name oleander comes from its superficial resemblance to the olive family, *Olea spp*.

As is common with many poisonous plants, it has become synonymous with death and ill fortune in many of the countries where it grows. In Tuscany and Sicily, the dead are covered with oleander blossoms before burial, and in India, the dead are crowned with chaplets of oleander during funerals.

It is also one of the most poisonous common garden plants. A member of the dogbane family, the entirety of the plant, including the smoke it produces when burned, is highly toxic. Although there have been only a handful of officially recorded deaths since 1985, just a few leaves are enough to kill a child. Ingestion causes abdominal pain, vomiting, rapid pulse and cardiac arrest, and simple contact with the plant can cause blistering and irritation. In India it is known as the horse killer, and in Italy ass-bane, as merely the scent of the flowers, or water that leaves have fallen into, are enough to kill domestic animals and livestock.

Various tales circulate of people having died from eating meat cooked on skewers of oleander wood. Some paint this as the fate of the Duke of Wellington's men during the Peninsula Wars, while in other versions it's unnamed Greek or Roman soldiers; sometimes it's hikers who fall foul of the error, or else unlucky boy scouts out camping. But the truth of these stories is as doubtful as the sources; oleander simply doesn't produce wood suitable for making skewers, and even if it did, it is highly unlikely that enough of the toxins would transfer themselves during the cooking process. Another similar legend of soldiers dying after sleeping on cut oleander branches is just one more example of fanciful conjecture.

However, there is a variety of oleander with a somewhat more insidious reputation. The yellow oleander (C*ascabela thevetia*), a relative of true oleander, is native to Mexico and *Central* America. The name *cascabela* comes from the Spanish *cascabel,* the rattlesnake's rattle—and the parallel between snake and plant is not without justification.

In the southern states of America, it is sometimes joked that old ladies who have tired of their husbands might use this plant to flavour a cake in order to do away with them, a quip which may be traced back to a famous ghost story from the Myrtles Plantation in St. Francisville, Louisiana. The story tells of Chloe, a slave owned by the family who ran the plantation, who for a brief while was taken as a mistress by the head of the family, Judge Woodruffe. Fearing that she would be sent back to the other plantation slaves once he tired of her, she determined to poison the family by baking oleander leaves into a birthday cake, after which she would ingratiate herself by nursing them back to health. Misjudging how many leaves to use, the outcome was instead that Woodruffe's wife and his two daughters died of as a result of the poisoning, and she fled the scene. The other slaves, not wanting to be thought part of her plan, caught and hanged her, and her ghost still haunts the plantation today.

While it's certainly possible to be caught out by the killing power of oleander—as such a tale illustrates—not all who ingest it do so accidentally or in ignorance of its potency. In Sri Lanka, the yellow oleander is called the suicide tree as it is a common method of such, particularly amongst the elderly, who often have access to the leaves and the seeds in care home gardens where it is commonly grown for decoration.

OTHALAM: Cerbera odollam

Still as
On windless nights
The moon-cast shadows are,
So still will be my heart when I
Am dead.

Adelaide Crapsey, *Moon-Shadows*

Whether it's called the pong-pong, mintolla, or the ominous 'suicide tree', the othalam tree is a common sight in its native India, most popularly grown as a hedge plant between home compounds.

A relative of the oleander and part of the lethal dogbane family, its scientific name is a reference to Cerberus, the guardian of the Greek Underworld. All parts of the plant, and even smoke from burning the wood, are toxic; but most of all so is the stone of the fruit. Referred to as othalanga, and no larger than an inch across, just one is enough to kill a human. The stone contains cerberin, which increases the potassium in the body's cells (the medical term for which is hyperkalemia, the same effect induced by lethal injection) and causes cardiac arrest, almost always fatally. Death after ingestion comes within one to two days.

The presence of cerberin is difficult to detect during autopsy, and the flavour of the stone is easily disguised in cooking, so othalam has become a common tool in both historical and present day suicides and murders. Between 1989 and 1999, 537 cases of othalam poisoning were confirmed in Kerala; approximately one per week in this one state alone, and one in five of all poisoning cases recorded in the same period.* The majority of these deaths were ruled as suicides, but thanks to the use of advanced autopsy techniques, the team involved in the

* Gaillard, Yvan, Ananthasankaran Krishnamoorthy, and Fabien Bevalot; *Cerbera Odollam: A 'suicide Tree' and Cause of Death in the State of Kerala, India*

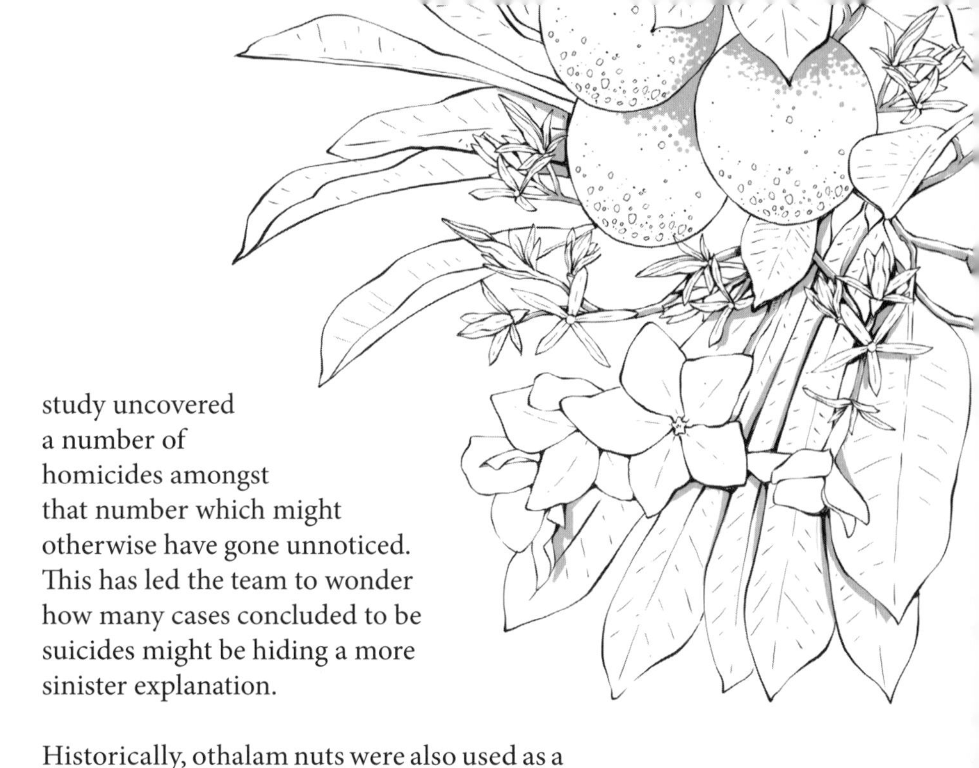

study uncovered a number of homicides amongst that number which might otherwise have gone unnoticed. This has led the team to wonder how many cases concluded to be suicides might be hiding a more sinister explanation.

Historically, othalam nuts were also used as a means of trial by ordeal. Most frequently recorded across Madagascar and Africa, this method of justice was usually reserved for serious accusations such as murder or witchcraft. The accused would ingest the nut of a poisonous plant, usually *C. odallam*, and if they vomited it back up without ill effect, they were declared innocent. If, however, they were unable to do so, they would be left to die of poisoning, or executed in a means befitting the crime. The popularity of this form of trial came from the belief that a good spirit presided in the plant that would strike only the hearts of the guilty, and so confident were they in the accuracy of the test that many would voluntarily take it to prove their innocence. Despite this, such ordeals were responsible for incredible numbers of deaths. On one occasion, over 6,000 people are recorded as having died in a single trial.*

In the case of a mutual dispute, both parties would undergo the ordeal, and whoever survived would be declared the innocent party. If

* Heiss, A., Maleissye, D., Tardieu, J., Viossat, V., Sahetchian, K.A. and Pitt, I.G.; *Reactions of primary and secondary butoxy radicals in oxygen at atmospheric pressure. International Journal of Chemical Kinetics,* 1991

they both survived, there were no grounds for the dispute, and if they both died, both were shown to have been dishonest in some form. If the individual who died was of a lower class, their body would be thrown to wild animals; however if they were of a higher class, their relatives would typically pay damages to the accuser. If they were unable to do so, they would sell themselves into servitude, usually to the winner. In some cases of the wealthy being accused, they might volunteer a slave or servant to undergo the ordeal for them.*

From a scientific standpoint, poison itself makes no distinction between the guilty and the innocent; however, it might have been possible to survive the ordeal by chewing and swallowing the nut quickly. This would cause a faster triggering of the vomiting reflex, thereby limiting the amount of toxin absorbed. A guilty party might chew the nut slower, fearing the outcome of the trial, and thus condemn themselves to death.

The method of trial by ordeal dates back to at least the 16th Century in Madagascar. On average, it is thought to have been responsible for the deaths of at least two percent of the population each year, until it was finally abolished in 1863 by King Radama II.† Outside of Madagascar, it is still practiced in Central Africa; however, it is reserved only for extreme cases due to the belief that an unnatural death is an offence against nature, and not to be inflicted without great consideration first.‡

There are only two other circumstances in which the ordeal would be undertaken with no crime having been committed. In West Africa, the intent is simply to induce vomiting and not death—for this, bark of the casca tree (*Erythrophleum guinenese*) is used, as it is high in tannins and produces vomiting before the plant's toxins can take effect. The ordeal is undergone by aspiring witchdoctors (the men who usually oversee the trials) who must submit to it several times before qualifying. It has also historically been used to anoint a new king, even when the title is hereditary; at least one of the prior king's sons must have submitted to

* Gwyn Campbell; *The State and Pre-Colonial Demographic History: The Case of Nineteenth Century Madagascar, Journal of African History*

† William Edward Cousins; *Madagascar of Today: A Sketch of the Island, with Chapters on its Past*

‡ Lasnet and Boye; *Poisons d'épreuve in Traite de Pathologie Exotique*

the ordeal twice with a willingness to undergo it a third time, or the throne would be declared vacant and open for contest.*

The majority of the plants used for these trials are thought to have come from the *Apocynaceae, Leguminosae,* and *Solanaceae* families. We know from descriptions of the poisons' effects that othalam, datura, and cassava were certainly amongst the most commonly used. Also common was the nut of the tanghin tree (*Cerbera tanghin*), a relative of othalam; casca bark (*Erythrophleum guinense*), so regularly used that it has come to be known as ordeal bark; and the poison bush, *Acokanthera oppositifolia.* This last was a popular choice, as its major toxins, ouabain (from the Somali *waabaayo*, meaning 'arrow poison') and strophanthin, are not always absorbed consistently by the digestive tract, which makes it impossible to predict a fatal dose. Therefore, a person might ingest a dose at one trial and live, only to ingest the same at a second trial and still die. This made it impossible for the accused to try and cheat the system by bribing the witchdoctors who oversaw distribution of the poisons.

Bribery was not an uncommon tactic for survival. Whether by means of money or nominating a slave to undergo the ordeal for them, it wasn't unheard of for defendants to do whatever they could to sway the odds in their favour. Corruption amongst witchdoctors also led to trials being used as a convenient way to be rid of high-ranking individuals. In one such case, a man who was not publicly liked was falsely accused of witchcraft. Unable to attend the trial as he was bedridden with illness, he was carried to the ordeal on his bed, and given a double dose of poison 'to first clear up the fever'.† In another case in 1881, an officer who was widely disliked was dealt with in a similar manner. Upon the death of his father, the officer was keeping the watch over the body—a common and perfectly innocent tradition of keeping a corpse company until burial. The village seized the opportunity to accuse him of necromancy, and at his trial prescribed him a heavy dose of poison to ensure that he would not survive.

* George Robb; *The Ordeal Poisons of Madagascar and Africa*

† Joannes Chatin; *Recherches pour servir à l'histoire botanique, chimique et physiologique du Tanguin de Madagascar*

PALA TREE: Alstonia scholaris

Said we, then—the two, then—"Ah, can it
Have been that the woodlandish ghouls—
The pitiful, the merciful ghouls—
To bar up our way and to ban it
From the secret that lies in these wolds—
From the thing that lies hidden in these wolds—
Had drawn up the spectre of a planet
From the limbo of lunary souls—
This sinfully scintillant planet
From the Hell of the planetary souls?"

Edgar Allen Poe, *To Ulalume: A Ballad*

The pala is one of the tallest trees in its native area, which spans India, Southeast Asia, and Australia. It can reach up to 140 feet tall, and grows fast, making it an ideal wood for pencils and blackboards, hence the specific epithet *scholaris* and the common name blackboard tree.

However, once darkness falls, it is also known as the ghost or devil's tree—and with good reason. A member of the dogbane family, its flowers bloom at night throughout November, and their pungent scent can cause severe headaches. In southern India, children are warned to stay away from this tree after dark, for it is said to be the haunt of the *yakshi*, a female vampire who preys on wealthy men and lures them to their death.

The yakshi is said to be the vengeful spirit of a woman who was the victim of a tragic love affair, and met her end at the hand of the man she was in love with. A beautiful woman who dresses in white, she waits under her tree to ask young men for a match to light a cigarette. When they are near enough, she transforms into a vampire and devours them, leaving only nails, teeth, and hair behind. In Malaysia, a similar creature called the *pontianak i*s said to haunt the same tree, but simply sucks the blood of her victims before releasing them.

Despite the yakshi's fearsome reputation, in northern India she (and her male counterpart, the *yaksha*) is described as a tree spirit, more peaceful than her southern counterpart. Rather than a man-eating vampire, she is a fertility spirit described as 'the fragrance in the blossoms of flowers', and can grant favours to humans who treat the tree well.

A theory regarding the aggression of the southern Indian yakshi is that it may actually be a pisacha (literally 'eater of raw flesh') in disguise. The pisacha is a demon is commonly associated with night-blooming trees, a vicious creature that haunts crossroads for its victims. However, iron will render them powerless, which provides an interesting connection with the Malay pontianak—in Malaysia it is believed that driving an iron nail into the neck of the creature will turn her back into a normal woman.

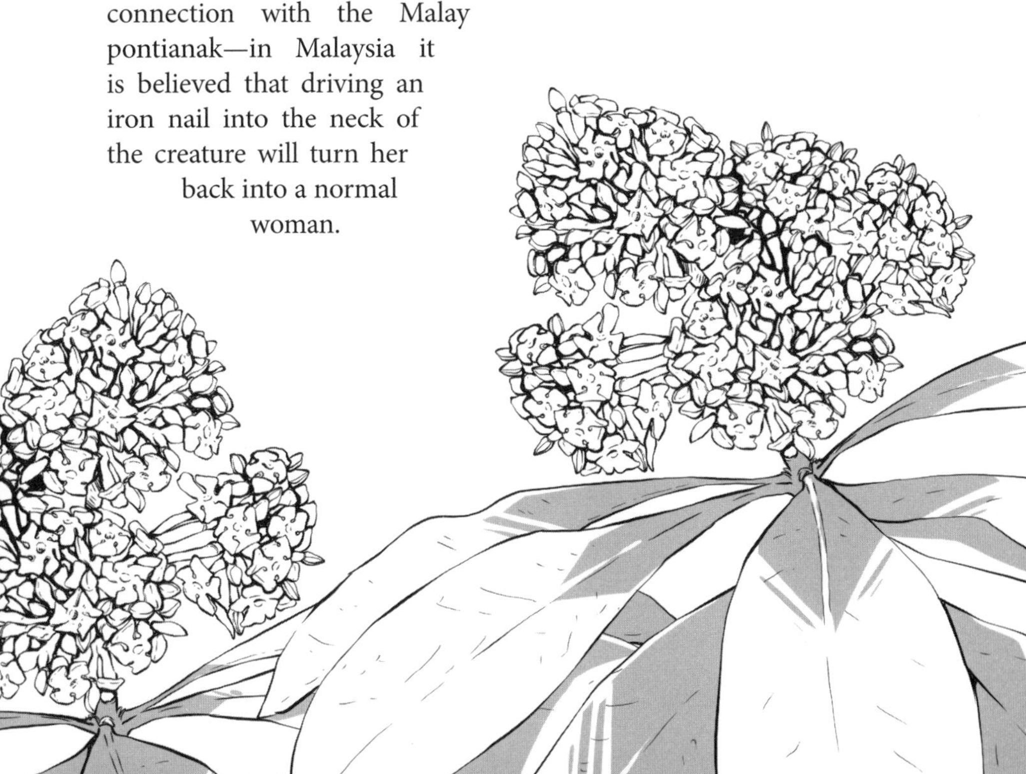

PEPPERS: Capsicum spp.

Out of a fired ship, which by no way
But drowning could be rescued from the flame,
Some men leap'd forth, and ever as they came
Near the foes' ships, did by their shot decay;
So all were lost, which in the ship were found,
They in the sea being burnt, they in the burnt ship drowned.

John Donne, *A Burnt Ship*

A cousin to the more infamous nightshade, capsicums are a member of the Solanaceae family and were domesticated at least 6,000 years ago. Their exact origins are difficult to pinpoint, but fruit, seeds, and pollen have been found at sites in the Tehuacan Valley in Puebla, Mexico dating to at least 4,000 BC, and in Huaca Prieta in Peru from around 2,000 BC.

The capsicum travelled beyond its native lands when it was discovered by early Spanish explorers, who found red pepper pods in the Caribbean while looking for the Brazilian peppertree (*Schinus terebinthifolia*). It's not known for certain whether their naming of their discovery as 'pepper' was due to a case of genuine misidentification, or whether these early explorers simply wished to save face over bringing back the wrong plant, but either way, the name stuck despite the lack of any connection between the capsicum and the peppertree.

Once introduced to Europe, the popularity of the plant took hold. Of all European countries, Hungary has become one of the most significant producers of capsicums and capsicum-derived spices (particularly paprika) on the continent. An old folk tale as to how the plant came to be cultivated Hungary goes as follows:

When Turkish soldiers invaded, they kidnapped a local girl and took her to live in their harem. The Turks ate a great many spicy foods, as it made them fierce in battle and hungry for women, and when the girl learned of this, she set about devising a way to return to her village and the boy she was engaged to. Upon her escape, she brought capsicum seeds home with her, and soon the plants grew across the country. The spice gave the Hungarian fighters the same strength as the Turks, and the invading force was defeated soon after.

Capsicum has no real benefit as an aphrodisiac, nor does it have the capacity to increase a man's strength, but it is well known for its burning heat. Caused by the compound capsaicin, the sensation doesn't actually cause burns but instead causes the nerves to send signals of a burning sensation to the brain. Though drinking water will do nothing to ease this, alcohol will dissolve the compound, and fats such as butter and milk will bind it. This survival technique likely evolved to deter mammals from eating the seeds, leaving them available instead for birds which swallow them whole and help to disperse them.

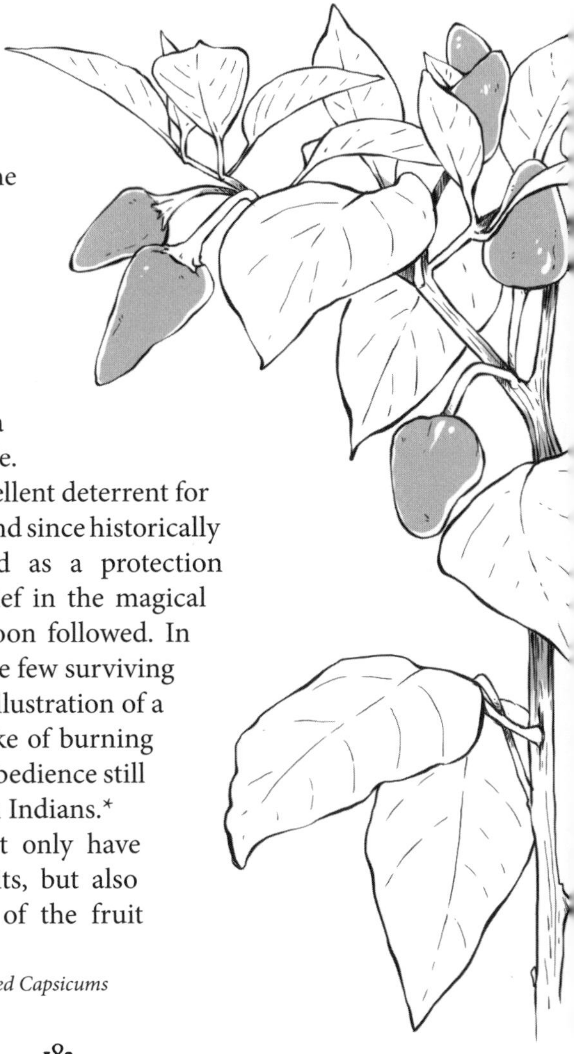

The properties of capsaicin have made chili peppers a popular fumigant worldwide. The irritating smoke is an excellent deterrent for rats, insects, and other pests, and since historically fumigation has been upheld as a protection against the supernatural, belief in the magical properties of chili peppers soon followed. In the *Codex Mendoza*, one of the few surviving Aztec books, there is also an illustration of a boy being held over the smoke of burning chilis—a punishment for disobedience still in use today by the Popolocan Indians.*

In Mexico, chilis not only have the ability to deter evil spirits, but also to kill them; the sweetness of the fruit

* Jean Andrews; *Peppers: The Domesticated Capsicums*

draws them in before destroying them with its fire. One such demon especially susceptible escape this fate is the vampiric *luban oko*, or 'red demon', which stalks the Tsachila or Colorados Indians of the Amazon. If a village is suspected of being haunted by this creature, villagers will burn chilis in a fire whilst also serving them in a feast. This way, the demon is foiled twice: he cannot eat the spicy meal, and the fumes from the fires will asphyxiate him.*

In Africa, pepper (particularly *C. frutescens,* the popular tabasco pepper) can be used to detect and trap witches. One such Ghanaian ritual involves lighting a fire under a tree where witches are thought to meet, and throwing dried peppers into the fire. The smell will entangle any witches present, and they will be unable to fly away.†

POISON IVY, OAK, AND SUMAC: Toxicodendron spp.

Now wound the path its dizzy ledge
Around a precipice's edge,
When lo! a wasted female form,
Blighted by wrath of sun and storm,
In tattered weeds and wild array,
Stood on a cliff beside the way,
And glancing round her restless eye,
Upon the wood, the rock, the sky,
Seemed naught to mark, yet all to spy.

Sir Walter Scott, *The Lady of the Lake*

Perhaps three of the most notorious plants in North America, tales are plentiful of those who have gone walking in the wilds only to fall foul of the *Toxicodendron* (previously *Rhus*) family. Poison ivy, oak, and sumac are all close siblings; ivy is *T. radicans* (rooting), oak is

* Juan Javier Rivera Andía; *Non-Humans in Amerindian South America: Ethnographies of Indigenous Cosmologies, Rituals and Songs*
† Hans Werner Debrunner; *Witchcraft in Ghana: A Study on the Belief in Destructive Witches and its Effect on the Akan Tribes*

T. pubescens (hairy, due to its hairy leaves), and sumac is *T. verix* (meaning resin, regarding its secretions).

Of the three, poison ivy is probably the best known, and certainly the most widespread. It was given the name by Captain John Smith of the English settlement at Jamestown, who described it thus: 'beinge touched, the poisoned weed causeth rednesse, itchynge, and lastly blisters'. As it grows in a manner similar to English ivy, it's easy to see where the epithet came from. Despite the similarity, however, it is actually a relative of cashews and pistachios, and bears no relation to English ivy.

But it wasn't just English settlers who hated the plant (a chronicler of Jamestown declared that 'there were never Englishmen left in a foreign country in such miserie as we were in this new discovered Virginia'). The native tribes harboured a healthy dislike of poison ivy too, although the Ramah Navajo and Cherokee alike used it as a dye plant and in the production of arrow poison.* Nevertheless, when forced to approach it, it was common practice to refer to it as 'my friend' in hopes of appeasing its difficult nature.†

For those who live in North America and encounter this plant regularly, it may seem incredible that it was, at one point in time, an eagerly sought-after commodity. But sought-out it was, by

* W. T. Gillis; *The systematics and ecology of poison-ivy and the poison-oaks*

† James Mooney; *History, Myths, and Sacred Formulas of the Cherokees, 1981*

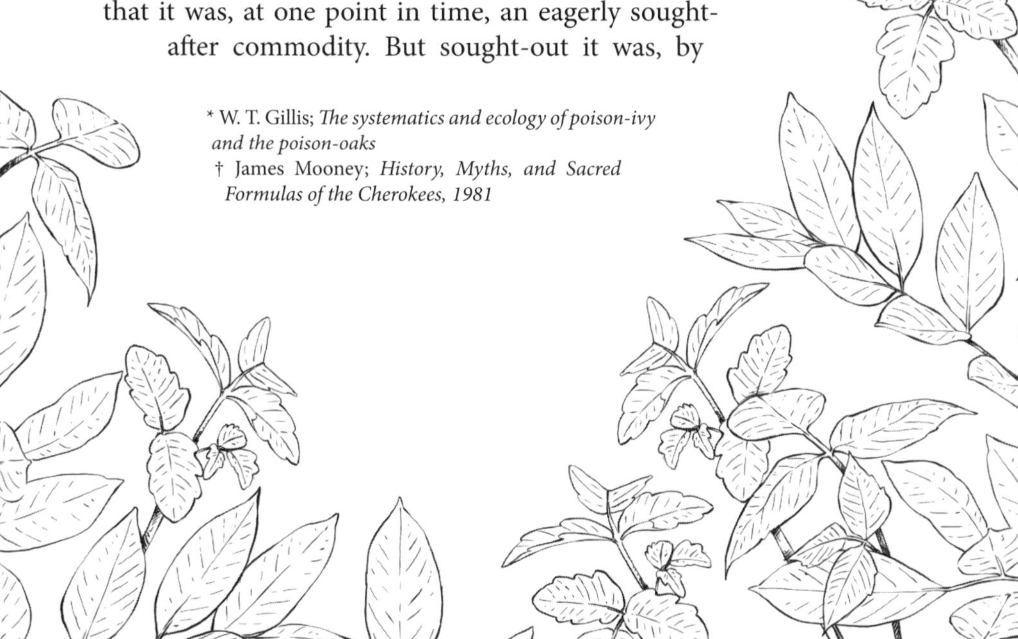

botanical institutions in Europe such as England's Kew Gardens or the Faculty of Medicine in Paris, which have curated collections of useful and unusual plants throughout the centuries: much of what we know about botany today comes as a result of their studies. In 1668, a sample of poison oak was sent by one Richard Stafford from Bermuda to England with a warning: 'I have seen a Man, who was so poyson'd with it, that the skin peel'd off his face, and yet the Man never touch'd it, onely look'd on it as he pass'd by'. A clear exaggeration, but with gruesome stories such as this attached to their reputations, it's no surprised that there was an avid (if brief) interest in studying these painful plants.

Yet despite their ill standing, poison ivy, oak, and sumac aren't technically poisonous. A poison is classed as a substance that causes illness or death when consumed or absorbed; but the *Toxicodendrons* excrete urushiol, an oil that activates the body's immune system, producing an allergic reaction in the form of a rash. And unlike many poisonous plants, this toxicity hasn't evolved as a defensive measure—in fact it is an unintended side effect, as the purpose of urushiol is to help the plant to retain water in dry periods. It is thought that approximately twenty-five percent of all people are immune to its effects, and so can handle the plants safely. For those who are susceptible, however, each subsequent exposure to the oil will get worse as the immune response intensifies. Luckily, a plant that contains urushiol can be easily identified—simply wrap a sheet of white paper around the stem or leaf, and crush the plant inside it. If the volatile oil is present, it will leave a brown mark on the paper.

Urushiol occurs in all members of the *Toxicodendron* family, and is named for the lacquer tree (*T. vernicifluum*), called *urushi* in its native Japan. When the sap of the tree—which contains urushiol—is exposed to air, it dries into the glossy, hard lacquer that is familiar on traditional Chinese, Japanese, and Korean lacquerware.

It was also made use of by Shugendō Buddhist monks in medieval Japan to mummify themselves alive. Known as *sokushinbutsu*, this difficult ten-year ritual is thought to have been attempted by many, but to date only 24 'successful' mummies have been discovered.

The process requires one thousand days of extreme fasting on a diet known as *mokujikigyo* ('eating a tree'): the participant relies only on pine needles, seeds, and resins found in the mountains for sustenance, which eliminates all bodily fat. This is followed by two thousand days of purging the body of all of its contents, and reducing liquid intake to dehydrate the body, drinking only a bowl of the lacquer oil every day, which renders the body too toxic for maggots and pests. Once the monk ceases drinking entirely, he enters a state of meditation, ringing a bell once a day to indicate that he is still alive. Only when the bell stops ringing is the tomb sealed for a final thousand days, where his remains become naturally mummified.

POPPY, OPIUM: Papaver somniferum

On the way, among the fields,
Poppies lift vermilion shields,
In whose hearts the golden Noon,
Murmuring her drowsy tune,
Rocks the sleepy bees that croon.

Madison Julius Cawein, *The Land of Hearts Made Whole*

Few things are more evocative of a lazy summer's afternoon than the sight of poppy heads nodding in the breeze. Common around the world, they are a familiar visitor to agricultural fields, parks, and roadside verges, providing a welcome splash of colour and a plentiful source of food for pollen-dependant insects such as bees, moths, and butterflies. Their propensity for growing in disturbed soil—such as that found on battlefields—has also led to the poppy becoming a symbol of remembrance for veterans and those who have died in conflict, the particular poppy chosen for the role being the corn poppy, *P. rhoeas*.

But the flower is also famous for its properties as a sedative and its role in the creation of opium. The poppy responsible for this particular association is *P. somniferum,* the breadseed poppy, also known colloquially as the opium poppy. So well-known was it for its ability to induce sleep that early Romans believed it created by Ceres who, seeking an escape from her grief at the loss of her daughter Proserpine, created the poppy as a means of granting herself oblivion. William Browne's *Consolation, Oblivion* speaks of this creation:

Sleep-bringing Poppy, by the plowman late,
Not without cause to Ceres consecrate.
Fairest Proserpine was rapt away,
And she in plaints the night, in tears the day,
Had long time spent: when no high power could give her
Any redress, the Poppy did relieve her:
For eating of the seeds, they sleep procured,
And so beguiled those griefs she endured.

Early doctors used it as a rudimentary medication for patients who required calming or sedation. In hotter climates, it is possible to grow more than one crop per year, making it more accessible than other soporifics such as mandrake, which takes three years to mature. The dried sepals, or the seeds themselves, could be mixed with tea to ease pain, coughs, and encourage sleep. Paul de Rapin described it thus:

The powerful seeds, when pressed, afford a juice
In med'cine famous, and of sovereign use,
Whether in tedious nights it charm to rest,
Or bind the stubborn cough and ease the lab'ring breast.

It wouldn't be until 1805 that morphine would be extracted from poppies, and as such opium—which is created from the dried latex of the seed capsules—was used as a painkiller and to soothe restless babies. Homer alluded to opium in his *Odyssey*, calling it *nepenthes*, the 'destroyer of grief'. He attributed its discovery to the Egyptians, and the substance is indeed mentioned as early as 1550BC in the Ebers Papyrus as a method for preventing 'the excessive crying of children'. Surprisingly, use of poppy by-products as a treatment for colicky babies has continued until as recently as the last fifty years.

From the 16th Century, opium was most commonly administered in the form of laudanum, a mix of opium and alcohol (usually brandy) that was prescribed as an analgesic. Usually taken by adults for pain and sleeplessness, it would also be spoon-fed to restless infants. Laudanum was later used in Godfrey's cordial, a medicine popular in the early 1700s which combined the painkiller with a sugary syrup as a sedative specifically for children. And for those who could not afford orthodox medicines, a solution noted in the 1800s in the fenlands area of Norfolk and Cambridgeshire (where opium addiction was a particular problem) was as follows: raw poppy seeds would be wrapped in muslin and soaked in tea or sugar water, then the bundle given to a sick child for them to suck on.

It may have brought a tired parent some brief respite, but the risks of using poppy derivatives to settle a child were severe, and the rates of child deaths from opium ingestion were high. In the 1600s, physician Nicholas Culpeper remarked on the trend: 'An overdose causes immoderate mirth or stupidity, redness of the face, swelling of the lips, relaxation of the joints, giddiness of the head, deep sleep, accompanied with turbulent dreams and convulsive starting, cold sweats, and frequently death.' With the potential to cause such effects, it's no shock that poppy tea and opium could also be turned to more nefarious purposes: in the 1800s, one might fall prey to the practice of 'hocusing', where political supporters the night before an election would spike the drinks of the opposition's voters with laudanum to ensure that they would sleep through the event.*

From Basingstoke, England, comes another tale which highlights a different danger of the poppy's sedative effects. It begins with a sign mounted in South View Cemetery: *Mrs Blunden, wife of William Blunden, was buried alive in this cemetery in July 1674. Parliament fined the town for its negligence.*

Alice Blunden was, as the sign says, buried alive—not once, but twice, before her untimely passing. While her husband was away on business, records state that Blunden drank such a quantity of poppy

* As mentioned in Charles Dickins' *Pickwick Papers* and Thomas de Quincey's *Three Memorable Murders*

tea that she fell into a deep sleep from which no one could wake her. A doctor attended the scene, and after holding a mirror to her face to test for breath, declared her dead. Her husband was contacted, and he requested that the funeral be delayed until he returned; however the summer was a hot one, and the rest of the family decided to go ahead with the burial to ensure that her body did not begin to rot. Alice was a large woman, and they had to prise her into the coffin, holding her arms and legs down while they sealed it.

Two days after the funeral, two boys playing in the cemetery heard a muffled voice and screaming coming from beneath the ground. They ran to their school, where they told several teachers, but they were known troublemakers and the headmaster punished them for lying. The following day, however, the headmaster went to investigate himself and heard the woman's pleas. It was evening by the time those with the authority to open the grave could be gathered, and when the coffin was open Alice Blunden practically sprung out of it, so tightly had her body been squeezed inside. She was found to be 'most lamentably beaten', which was thought to be the result of the tight fit, and when they examined her they found no further signs of life present.

With nowhere to store the corpse until the coroner could be summoned the next day, they returned her to the grave overnight. As the grave was not fully covered, they assigned a guard to stand watch; but that night the weather turned to heavy rain and the guard retreated from his post, leaving the grave untended. Come morning, the party and the coroner 'found she had torn off a great part of her winding sheet, scratched herself in several places, and beaten her mouth so long till it was all in gore blood.'*

Alice was this time, at least, definitely dead. As no one person could be found responsible for the tragedy, the town was fined for negligence, and the method of checking for death using a mirror, though a recognised technique at the time, was retired from recommended practice.

Many of the mythological and superstitious connections of the poppy are entwined with its somniferous and quieting effects—or, in

* Francis Baigent and James Millard; *A History of the Ancient Town and Manor of Basingstoke*

the wrong hands, its ability to kill. The Greco-Roman gods Hypnos (Sleep), Thanatos (Death), and Nyx (Night), were often shown holding the flowers or else crowned with them, and poppy seeds would be scattered inside of the coffin at a funeral to encourage the body to sleep rather than return to life.

The use of poppy seeds in this manner may relate less to their soporific properties, however, and more to an older, traditional ward against demons and vampires. The superstition in question travelled from the Roman Empire across Europe, and claims that, should something that is small, numerous, and easily scattered be thrown into the path of a monster, it will be compelled to stop and count each tiny piece before continuing, allowing you time to escape pursuit. This myth has been retold in various forms, and besides poppy seeds also names rice, breadcrumbs, and acorns (amongst others) as suitable distractions. You may even keep witches or any other visiting evil from your door by growing a rosemary bush beside it, so that—by the same logic—they will be compelled to stop and count the numerous tiny leaves. Perhaps it is in this context, then, that substances such as poppy seeds have been found in ancient burial sites: a deterrent to prevent a corpse from being preyed upon by these creatures… or from turning into one itself.

RHUBARB: Rheum rhabarbarum

Volga, Volga, Mother Volga,
Wide and deep beneath the sun,
You have ne'er seen such a present
From the Cossacks of the Don!
So that peace may reign for ever
In this band so free and brave,
Volga, Volga, Mother Volga,
Make this lovely girl a grave!

Dmitri Sadovnikov, *Stenka Razin*

There's nothing quite like hot, stewed rhubarb with a dash of ginger. A favourite of gardeners and foodies alike, this tart vegetable grows the world over and is near impossible to kill. But it isn't native to most of the countries that consume it, and it's responsible for more than just a few deaths—so what exactly is its story?

Rhubarb originally hails from Mongolia and northwest China, where the roots were used for medicinal purposes and were highly praised for their purgative properties as far back as 2,700BC. Marco Polo, on his famous journey to Cathay, remarked that rhubarb grew 'in great abundance' in these areas, and it became a major export of the Silk Road to Europe, where it commanded a higher price than cinnamon, saffron, and opium.

Along the Silk Road it made its way to Russia, where culinary varieties were first cultivated, and it is from here that we can trace the origin of its scientific name. Originally *Rheon rhabarbaron,* meaning 'from the barbarous lands of Rha' (Rha being the Greek name for the Volga River, which runs through much of Russia), the name eventually became convoluted and took the form we use now: *Rheum rhabarbarum.* Rhubarb continues to be an important plant in Russia, and even the infamous Russian witches, the Baba Yaga, are said to gather rhubarb from the banks of the Volga for use in their spells.

In 2014, a New York construction site clearing the grounds of a former German beer garden unearthed a 200-year-old glass bottle labelled 'elixir of long life' alongside bottles for other medicinal bitters. Alcoholic cure-alls experienced a massive surge in popularity in the 19th Century, and were commonly available at most bars. The recipe for the elixir discovered on the construction site was tracked down in a 19th Century German medical guide, and included rhubarb (likely the root, or as a flavouring), gentian root, turmeric, saffron, aloe juice, and grain alcohol. Most of these ingredients encourage good digestive health and genuinely may have contributed to a healthy lifestyle, although any immediate sprightliness felt by the drinker was more likely to do with the alcohol content.

But while a swig of something with rhubarb root or stalk in it might be all well and good, the leaves—and some questionable advice from the government—were directly responsible for the deaths of eighteen people over the course of World War I and II. During these periods, civilians in Britain were encouraged to embrace

the 'Dig for Victory' ethos—using what spare land they had in their gardens they were advised to keep livestock or grow crops, and the government distributed hundreds of pamphlets with information on the best ways in which to make the most of this home-grown bounty.

All was well until a pamphlet was released containing advice from Maud Grieve, a leading botanist in the early 1900s who was hugely popular despite plentiful errors in much of her writing. Thanks to her guidance, this pamphlet listed rhubarb leaves as a vegetable—misinformation likely taken from a letter published in *The Gardener's Chronicle* in 1846, in which the gardener for the Earl of Shrewsbury remarked upon the edibility of the entire plant.

Unfortunately, this gardener was incorrect; the leaves contain oxalic acid, and though large amounts of the leaves have to be eaten in order to be fatal, death is preceded by vomiting, convulsions, nosebleeds, and internal bleeding, all within just one hour of ingestion. The acid combines with calcium in the body to rapidly form kidney stones, and ultimately causes kidney failure. And while the gardener's misinformation was corrected in a later issue of the magazine, it is likely that many who read that first article may not have seen the amendment.

Upon this advice, the government promoted rhubarb leaves as a replacement for cabbage and spinach. Thirteen people died from rhubarb poisoning in the First World War, and the pamphlets were rapidly recalled when the error was discovered. Recalled, but not destroyed, for during the Second World War some industrious recycler found them in storage and ordered them to be redistributed, leading to a further five deaths.

Despite the danger of the leaves, rhubarb remained hugely popular in Britain across the course of both wars. It easily absorbs the flavour of other fruits when they are cooked together, and in a time of strict food rationing, many jams and marmalades were bulked out with a high rhubarb content. A popular urban legend is that one company made a huge profit by using exactly this method, but concealed the fact that their raspberry jams were up to eighty percent rhubarb by creating wooden raspberry seeds, thereby making them look more authentic.

ROSARY PEA: Abrus precatorius

Alone, alone,
Upon a mossy stone,
She sits and reckons up the dead and gone
With the last leaves for a rosary,
Whilst all the wither'd world looks drearily,
Like a dim picture of the drownèd past
In the hush'd mind's mysterious far away,
Doubtful what ghostly thing will steal the last
Into that distance, gray upon the gray.

Thomas Hood, *Autumn*

Also commonly called the jequirity bean, *A. precatorius* is an invasive tropical plant that has spread from its native Asia and Australia to make a tenacious pest of itself in the Caribbean and warmer American states. With deep-growing roots and vines that spread rapidly, it is quick to appear and hard to remove. A distinctive plant, it sports clusters of violet flowers and, for several months a year, pods packed with tiny red and black spotted seeds. The seeds retain their bright colours once dried, and are best known for their use in souvenir jewellery.

Despite its invasive nature, it has its uses. The seeds weigh approximately one carat, and are so reliably sized that they have been used for centuries in India as weights, named *rati,* particularly for weighing gold. They were also the basic unit in the Akan weighing system, in which ten seeds are the equivalent of the smallest brass weight, known as *ntoka.** In India and Java, the roots and leaves are prized for their liquorice flavour, and are commonly used as a substitute for such; it is sometimes known as Indian or Wild liquorice, and in Jamaica the name is shortened to the colloquial 'lick', or 'lickweed'.†

And yet, the rosary pea is one of the world's most poisonous plants. Ingestion of a single seed (weighing no more than 0.2 grams)

* Margaret Webster Plass; *African Miniatures: the Goldweights of the Ashganti*
† Martha Beckwith; *Notes on Jamaican Ethnobotany*

can be fatal to a fully-grown adult, and sale of these seeds is highly restricted in the UK under the Terrorism Act. Its toxin, abrin, is 75 times more powerful than the deadly castor oil plant, and causes vomiting, convulsions, liver failure, and death. Fatalities usually come from ingesting the seeds, but the toxin can also be absorbed through the skin, as well as inhaled from crushed seeds, or from drinking water that the seeds have soaked in. In India, where the plant grows rife in rural villages, the crushed seeds are often reported as the cause of death in suicides.*

In 2011, jewellery made from the seeds was subject to a mass recall from tourist locations across the UK, one year after an English woman suffered the full effects of wearing such an accessory. Having bought a bracelet made of the beans online, she began to suffer hives, mouth ulcers, vomiting, and hallucinations. With doctors unable to find the cause, she was sectioned under the Mental Health Act, losing her job and her home in the process. It was only when her son's school sent out a warning about these bracelets that she ceased wearing it, and her health rapidly returned. If the seeds can have such harmful effects through skin contact alone, it is no surprise that, in many of the countries where the plant grows natively, it is believed to house an evil spirit.

* Aishwarya Karthikeyan and S. Deepak Amalnath; *Abrus precatorius Poisoning: A Retrospective Study of 112 Patients*

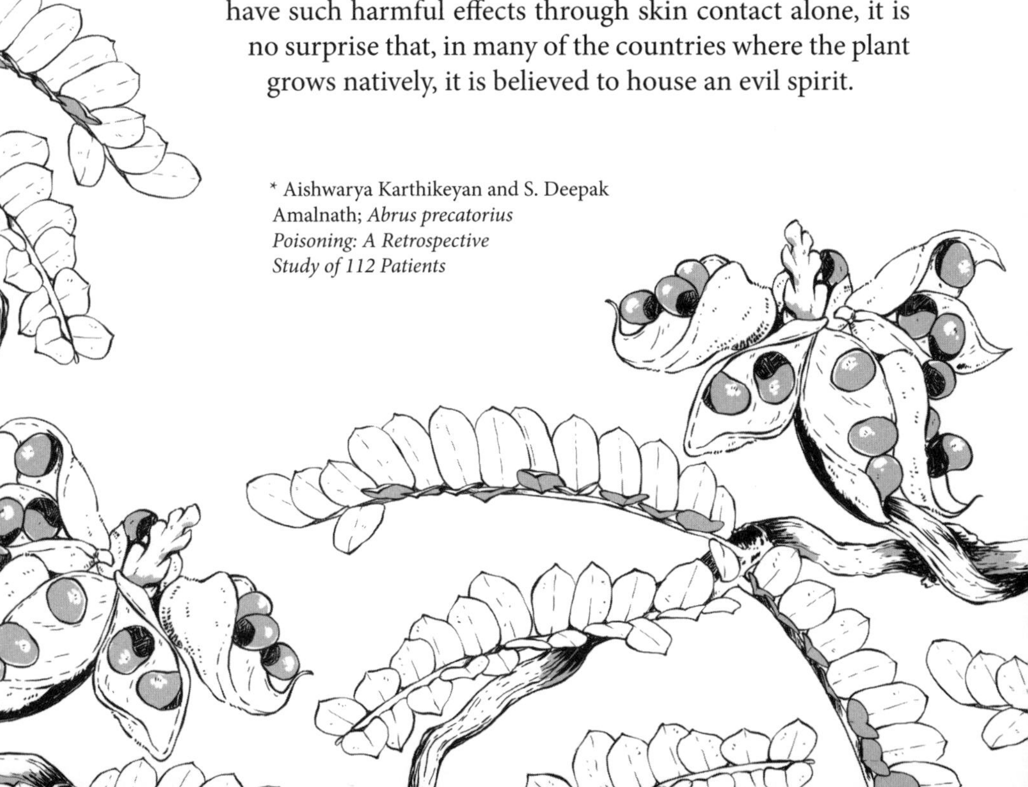

In southern Africa, the beans are associated with dangerous magic and sorcery. They are used solely for the decoration of items used in magical ceremonies, and worn only by witchdoctors. In India, too, it is associated with magic, and is dedicated to Indra, the supreme god of the Vedic Olympus. Indra fulfils a similar role to other Indo-European deities such as Zeus and Thor, being the god of the heavens, thunder, storms, and war. The roots of the plant are used in a ceremony intended to prophesise future events. Crushed together with the roots of the chaff flower (*Achyranthes aspera*) and black pigweed (*Trianthema decandra*), and mixed with castor oil and soot, the concoction is then applied to the palm of a child who, 'looking upon it, details what he perceives in the talismanic mixture, when accompanied by certain words of power, acting as a magic mirror, in which strange things become distinctly visible.'*

In the West Indies, the seeds are more benignly regarded and are mostly used as ornamental beadwork and as prayer beads (the Scientific *precatorius* comes directly from the Latin *precari*, to pray, as does the common name rosary pea). There seems to be no fear here about the toxic properties of the seeds, and they are often strung into bracelets and worn around the wrist or ankle to ward off evil spirits. Green, black, and white varieties of the seeds are used to create vibrant designs.†

* Qanoon-e-Islam, Jaffur Shurreef; *Customs of the Moosulmans of India, translated by Gerhard Andreas Herklots*
† Gooding, Loveless, and Proctor; *Flora of Barbados*

ROSE: Rosa spp.

Tenterhook among woods the spiteful briar is, burn him that is so keen and green;
He cuts, he flays the foot, he who would advance he forcibly drags backward.

Unknown author, *The Violent Death of Fergus mac Léti*

Ask anyone in the English-speaking world what they associate with roses, and most answers are likely to be the same: romance, marriage, and beauty. But look back a few centuries, and you'll find that this beloved flower also has a darker side: it was once symbolic of secrets, of death, and of sorcery. The briar is even known in France as *Rose Sorciere*, the Sorcerer's Rose, as it was supposed to have been planted by the Devil in a failed attempt to create a ladder back to Heaven.*

But it's the more appealing associations that remain today. One of the most widely grown and gifted flowers in the world, roses have been cultivated since at least 500BC, and were avidly bred from their wild cousins by the Persians, Egyptians, and Chinese into the garden varieties that we know now. The flowers are also associated with the Virgin Mary in Christian tradition; the word 'rosary' even originates from *rosarium,* meaning a garland of roses, as it is believed that early rosaries may have been made from strings of rosehips.

The connection with romance is probably as old as roses themselves. Buildings during Roman weddings would be crowned with roses, and the legends about the epic love between Cleopatra and Mark Anthony claim that she once filled her bed chamber two feet deep with the petals in order to seduce him. A common motif in many traditional Scottish and English ballads is that of two plants springing from the graves of tragic lovers who died young; the plants typically end up twining over the space that divides the graves, so that the lovers

* Paul Sébillot, *Le Folk-Lore De France: La Faune Et La Flore*

may be together in death. This motif appears in the medieval legend of Tristan and Isolde, whose graves are united by an ivy bush, and takes the form of roses in other famous songs such as Barbara Allen, Lord Love, and Lady Margret:

Barbara Allen was buried in the olde church yard,
Sweet William buried beside her.
Out of Sweet William's heart grew a red, red rose,
Out of Barbry Allen's, a briar.
They grew and grew in the olde church yard,
'Til they could grow no higher.
At the end they formed a true lover's knot,
And the Rose grew 'round the Briar.

The domestic rose's untamed cousins, the wild (*R. rubignosa*) and dog (*R. canina*) roses, have the Romans to thank for their historical significance. These are the varieties that the Romans would have been most familiar with, and their early reputations were spread as the Empire grew throughout Europe.

Prior to the advent of Christianity, early Romans celebrated the feast of *Rosaria* (or *Rosalia)*, the Festival of Roses. After Christianity took over as the main Roman religion, Rosaria was converted into the Christian Pentecost, also called the Easter of Roses. Celebrated on dates throughout May to July, this event was a commemoration of the dead, during which graves would be tended and the dead remembered. Roses would be offered to the *manes,* the souls of the deceased, who were thought to become protective deities of the home after passing. Military legions would make sacrifices and crown their standards with flowers, and offerings would be left at temples and statues. Bloodless sacrifices, too, would be made, not only in the form of wine but also that of roses and violets; the colour of these latter together intended to represent the colours of blood and the rot of death.

To the Romans, roses were the favoured flower for funerals and memorials; they would adorn the tables at funerary banquets, and were a frequent sight on monuments. A symbol of both mourning and beautiful youth, they were especially associated with the deaths of young people, and funerary epitaphs commonly referenced the concept of the bodies becoming flowers after death. One such Latin inscription reads:

*Here lies Optatus, a child ennobled by devotion: I pray that his ashes may be violets and roses, and I ask that the Earth, who is his mother now, be light upon him, for the boy's life was a burden to no one.**

The Greeks shared many of these traditions; they carved roses on funerary steles, and crowned the dead with chaplets of roses. In Homer's *Iliad,* Aphrodite anointed Hector's corpse with rose oil to ensure that his body would always remain fresh—a practice also used by ancient Egyptians during the embalming process.

The Romans also imbued the rose—most notably the briar—with a connection to secrecy and hidden treasures. It's an association that can be found throughout fairy tales and folklore: think of the Arthurian tower of roses in Broceliande forest that trapped Merlin within its walls, or the briars that hid away Sleeping Beauty. The Greeks tell of Eros, who gifted a rose to Harpocrates, the god of Silence, to ensure that his mother's indiscretions would remain secret.

* Jocelyn Toynbee; *Death and Burial in the Roman World*

These associations likely came from the Latin phrase *sub rosa,* or 'beneath the rose'. In Roman dining rooms, if the room was to be privy to conversations that must remain in absolute confidence, a rose would be hung from the ceiling to indicate that participants must keep events strictly to themselves. Confessional booths in many Christian churches are carved with roses for this same reason, and it was due to this connection that the white rose became a symbol of the Jacobite rebellion in 18th Century Scotland. The Scottish government still uses the notation 'sub rosa' today when discussing confidential issues.

In England, the rose—specifically the Tudor rose—appears as an heraldic emblem on the royal coat of arms of the UK. The Tudor Rose is a combination of a white rose and a red. These were the emblems of the Houses of Lancaster and York, two rival branches of the Royal House of Plantagenet. The opposition between these houses is infamous, but most notably came to a head in a number of civil wars spanning 32 years during the 1400s known as the Wars of the Roses. After the houses finally united to form the House of Tudor, the Tudor Rose was created to symbolise the union. There still exists now a cultivar known as the York and Lancaster rose, which is said to have originally grown from the blood on the battlefields where the two sides clashed.

SILK COTTON TREE: Ceiba pentandra

Lo her growths of sons, foliage of men and frondage,
Broad boughs of the old world tree,
With iron of shame and with pruning-hooks of bondage
They are shorn from sea to sea.
Lo, I set wings to thy feet that have been wingless,
Till the utter race be run;
Till the priestless temples cry to the thrones made kingless,
Are we not also undone?

Algernon Charles Swinburne, *Quia Multum Amavit*

One of the tallest trees in Central America, the silk cotton tree, also called the ceiba, can grow up to 200 feet tall—only slightly shorter than some of California's famous sequoias. To the ancient Mayan peoples, it was sacred, and not to be cut down; they called it *Yax Che*, the First Tree, and according to their mythos it was the symbol of the universe and the centre of the earth. The Maya saw the world as a quincunx, consisting of four directional quadrants and a central space corresponding to the fifth direction, occupied by the silk cotton tree. As with many occurrences of the World Tree in mythology, the roots of the tree were believed to grow down into the underworld, and the spreading branches to reach up into the heavens (of which the Mayans had thirteen). The trunk represented the terrestrial world where humans live, able to travel up

and down at
the beginning and end of their lives.*
It's no surprise, then, that the Mayans—
and modern-day tribes that live along the Amazon River—revered this tree so much. In addition to its towering height, its canopy can be as wide as 140 feet—almost as wide as it is tall—and the cotton-like fluff that grows in its seedpods can be spun into fibre. The resulting product is light, resilient, and resistant to water, making it perfect for insulation, stuffing, and for wrapping around blowgun darts, creating a seal that aids in the propulsion of the dart through the tube.

It is also useful to the local ecosystem, as its night-blooming flowers providing an important source of pollen for nocturnal insects and bats. The Tainos, an indigenous people of Jamaica, believe that the forest is inhabited by *opias*, spirits of the dead. They can be identified by their lack of navel, and come out at night to feed on guayaba fruit and ceiba flowers. The fruit- and flower-eating bats of the forest are

* Timothy Knowlton and Gabrielle Vail; *Hybrid Cosmologies in Mesoamerica: A Reevaluation of the Yax Cheel Cab, a Maya World Tree*

thought to be the physical forms of the *opias*.

The night-blooming natures of many flowering trees have associated them in local lore with stories of vengeful creatures and female ghosts, and the ceiba is no different. The Maya believed that the *xtabay*, a malign female being, hid within its trunk.* It may seem strange for a tree so central to the Mayan faith to be inhabited by a bad spirit, but the name is thought to have evolved from Ixtab, the Mayan goddess of suicide and the gallows. In Mayan culture, the act of suicide, particularly by hanging, was considered an honourable way to die, and those who did so in this manner would be retrieved by Ixtab and taken to paradise.

Today, the xtabay still exists in some forms. The soucouyant of Haitian, Louisianan, and Caribbean lore is a creature which also resides within the ceiba. By day, she appears as an old woman, but once the sun has set she sheds her skin and turns into a fireball that can enter homes at will and drain the blood of the occupants. Like most vampiric cryptids, she has a compulsion to count items and can be slowed down or avoided by scattering grains. But unlike other vampires, who suck blood for sustenance, the soucouyant takes this blood back to the ceiba tree, and trades it for favours from other demons that also live there. It is unknown what these demons do with the blood that they gather, but due to their presence, the tree is also known as the Castle of the Devil in the islands of Trinidad and Tobago.

In the Caribbean, the ceiba is instead a magnet to duppies and jumbies, ghosts who roam the earth after dark. Jumbies are always malevolent, but duppies can be more gentle; their personality depends on the person they were during life. Due to the spirits that it attracts, it is bad luck to cut down a silk cotton tree, as doing so will free the spirits from its thrall, bringing misfortune to anyone living nearby.†

* Jesus Azcorra Alejos; *Diez Leyendas Mayas*

† Zora Neale Hurston; *Tell My Horse: Vodoo and Life in Haiti and Jamaica*

SOLANACEAE: Solanaceae spp.

Fine to taste they are,
Smoothly globular,
Fed by the sweet brook
In their shady nook.
Fronds at top and toe
Clutch them round, as though
They are hearts of sheep
In the eagle's grip.

Ibn Sara of Santarem, *Aubergine*, circa 1123

There are a number of Solanaceae relatives with individual entries in this book, such as the mandrake, bittersweet, brugmansia, and of course the infamous deadly nightshade; but this family is vast and full of plants worthy of at least a passing mention. Many cousins of the aforementioned are, surprisingly, found in our salad bowls and gardens; but even some parts of these edible plants—usually the leaves and stems—can be just as toxic as their deadlier relatives.

Most members of the Solanaceae family contain solanine, which in small doses has a narcotic effect, and in larger amounts can be responsible for convulsions and death. The scientific name *Solanum* is thought to come from the Latin *solamun,* meaning 'to comfort', or *solare,* 'to soothe'. A European cure for restless babies once involved putting the leaves of black nightshade (*Solanum nigrum)* in their cradles,* and amongst some South American peoples, leaves of the currant tomato (*Solanum pimpinellifolium,* the wild ancestor of the cultivated tomatoes we're more familiar with) are steeped in water as a cure for insomnia.†

AUBERGINE: Solanum melongena

Aubergines were first cultivated in China around 544AD, coming to the wider world a couple of centuries later and achieving widespread

* Maude Grieve; *A Modern Herbal*

† Michael Weiner; *Earth medicine – Earth Foods: Plant Remedies, Drugs and Natural Foods of the North American Indians*

popularity around regions of the Middle East and the Mediterranean in particular. Because of their rapid international expansion, they are commonly known by no less than six completely disparate names across the globe today, the most notable being aubergine and eggplant. Aubergine is the modern British name and comes from the Spanish *alberengena* (derived from the Arabic *al-bādhingiān*), while the preferred American name, eggplant, did not exist until 1767 when a white variety with egg-shaped fruits was cultivated.

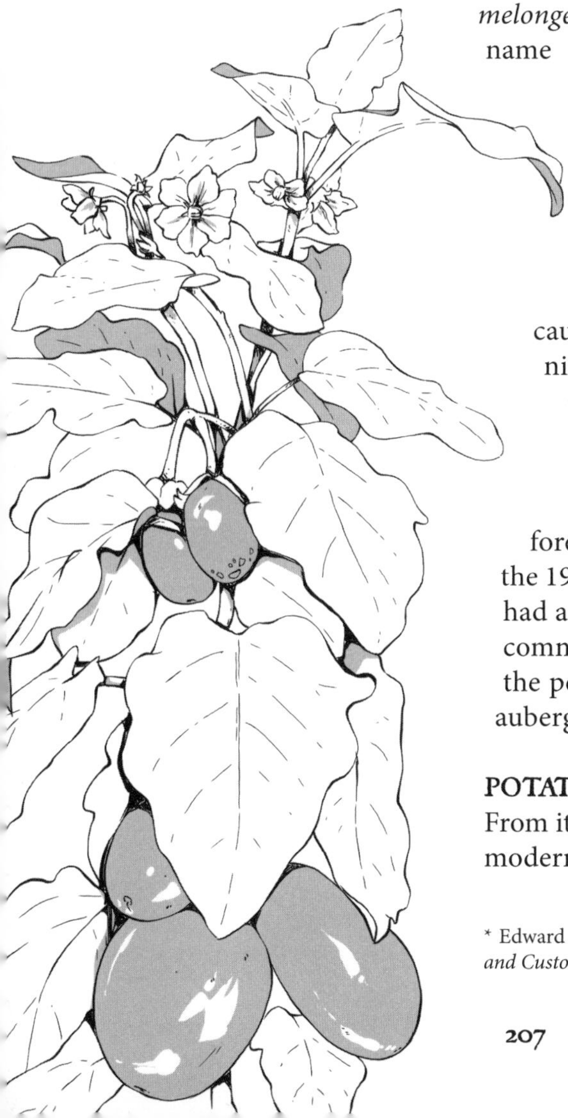

The scientific epithet *melongena* has roots in the Italian name *malanzana,* which is a corruption of *mela insano,* 'mad apple'. Though the toxin solanum can be found in the leaves of the plant, there is no proof that consuming aubergine—fruit or otherwise—will cause madness. However, the nickname has stuck. John Gerard's *Great Herball* says that 'doubtless these Apples have a mischievous qualitie, the use whereof is utterly to bee foresaken', and even in as late as the 19th Century, modern Egyptians had a saying that insanity was 'more common and more violent' during the period in the summer when the aubergine bears fruit. *

POTATO: Solanum tuberosum

From its humble roots in the region of modern-day Peru, the potato—which

* Edward William Lane; *An Account of the Manners and Customs of the Modern Egyptians*

was first cultivated around 8,000BC—has become an important crop in many countries around the world. First reaching European shores in the 1500s, they even enjoyed a brief period of glamour in the 18th Century: Marie Antoinette liked the flowers so much that she wore them in her hair, and the blooms became a brief fashion statement amongst the French aristocracy. By the 1850s, the potato had become—and still remains—the world's fourth largest food crop, after rice, wheat, and maize.

And, as with any widely-grown staple, tales of mystery and monsters have sprung up wherever this humble tuber is to be found. In Germany, precautions were once taken against blight from the *Kartoffelwolf* (literally, 'potato wolf'), which is said to lay in the soil and wait for the new year's crop of potatoes, eating half and spoiling the rest. For some time Germans also believed that potatoes, when left to rot, emitted a light bright enough that someone could read a book by it. An officer at the barracks in Strasburg reported thinking that the officer's quarters were on fire due to the fierceness of the light coming from a cellar full of ageing potatoes.*

But like other members of the Solanaceae family, potatoes contain solanine, and another glycoalkaloid known as chaconine. The presence of these toxins in wild potatoes is concentrated enough to cause detrimental effects to humans, though in modern cultivars they have mostly been bred out and occur only in the green parts of the plant and fruit. They affect the nervous system, causing headaches, confusion, digestive distress, and, in severe cases, death. Though cooking usually destroys any toxicity, the compounds increase in concentration as they age, and old potatoes can contain up to 1000mg per kilogram of solanine, many times the recommended maximum amount that is safe for ingestion.

TOMATO: Solanum lycopersicum

Common though it may be in our kitchens, the tomato is a fruit with a rather dubious reputation. Like the potato, tomatoes originated in South America and came to Europe in the early 16th Century, and as with other relatives in this family, the leaves and stems contain

* Richard Folkard; *Plant Lore, Legends, and Lyrics*

solanine. The leaves, when made into a tea, have been responsible for at least one recorded death.*

When the tomato first arrived in Europe around 1540, it did so in the middle of a continent-wide witchcraft panic. The variety of tomato that was introduced was similar to the yellow cherry tomatoes that we know now, and to the untrained eye, appeared to be the same thing as nightshade or mandrake, both botanical relatives. Around this time, tales of salves that could make you fly and ointments that could transform a person into a wolf ran rampant amongst peasantry and nobility alike, and anything unfamiliar—especially something imported from unknown, foreign lands such as the Americas—was under instant suspicion.

Eager to understand their magical foes, werewolf and witch hunters turned to old manuscripts believed to contain information on the occult. Many of these suspected books on magic—like the treatises produced by Galen of Pergamon, one of the most prolific Greek medical researchers—included in their contents descriptions of plants or animals as yet unnamed or of uncertain identity. These new, mysterious American imports were scrutinised to see whether they corresponded to any such gaps left in these texts; and unfortunately, the common tomato seemed to fit exactly one such description.

Galen's writings mention in some detail a plant written as λυκοπέρσιον, a half-word of which only the first part, 'wolf', can be understood. Transliterated as *lycopersion,* this became mis-transcribed during the 16th Century as *lycopersicon,* or the 'wolf peach'. Galen's description spoke of a poisonous Egyptian plant with golden fruits, a ribbed stalk, and a strong smell, and even as early as 1561 there was speculation between Spanish and Italian botanists that the wolf peach might, in fact, be a tomato. Though traders knew the tomato originated from the Andes, not Egypt, the controversial classification proved hard to discredit. Even Joseph Pitton de Tournefort, Louis XIV's personal botanist, went so far as to endorse this misconception in his hugely influential *Elemens de Botanique,* calling the tomato the *lycopersicum rubro non striato*—the ribless red wolf's peach.

* D. G Barceloux; *Potatoes, Tomatoes, and Solanine Toxicity (Solanum tuberosum L., Solanum lycopersicum L.)*

It didn't help matters that, for a short time, the tomato also earned the nickname of 'the poison apple', as a remarkable number of aristocrats seemed to sicken and die after eating them. The truth of the matter was that most of the plates used by nobility at the time were made of pewter, which has a high lead content. The acidic nature of tomatoes meant that when they were sliced on the plates, the fruit would leach the lead from the pewter, resulting in numerous fatal cases of lead poisoning. Once again, the innocent tomato was framed for the deed.

STRANGLER FIG: Ficus spp

Of later date of wives hath he read,
That some have slain their husbands in their bed,
And let their lover ride them all the night
While that the corpse lay on the floor upright:
And some have driven nails into their brain,
While that they slept, and thus they have them slain:
Some have them given poison in their drink:
He spake more harm than hearte may bethink.

The Wife of Bath's Tale, Geoffrey Chaucer

The name strangler fig, also called banyan, is a broad term given to any number of fig species that begin their life as an epiphyte: a plant that grows on another plant. Although the term has broadened to encompass other fig species, the name 'banyan' was originally given to *F. benghalensis*, the Bengal fig, which is the national tree of India. It comes from the Gujarati *baniya*, meaning 'merchant', as traders were seen to frequently rest and set up their stalls beneath the shade of these trees during hot weather. Portuguese traders misunderstood this word to refer specifically to Hindu merchants, and the word was adopted by the English in 1599 with the same meaning. By the early 1600s, these trees that provided the shade were being referred to by English writers as the 'banyan tree'.

The alternate name, strangler fig, comes from the growth pattern of this plant. Banyans are epiphytes, which means that the seeds, spread by birds, often germinate in the canopies of other trees and begin to grow without ever touching the ground. Growing downwards, they engulf the host tree in a cage of roots, until eventually the host is smothered and rots away, leaving a hollow mesh of banyan vines which eventually thicken into trunks. In the case of very old banyans, these roots can spread over such a huge area as to give the appearance of an entire grove of trees, each one of them connected directly to the primary trunk.

A particularly impressive specimen in this regard exists in Anantapur, in the state of Andhra Pradesh in India, and is known locally as Thimmamma Marrimanu (Thimmamma's Banyan Tree). The Thimmamma Marrimanu is more than 550 years old, and is believed to be the largest tree in the world; its canopy is over 19,000 square metres, and its branches spread over eight acres of land. According to local lore, the tree came into existence in 1434, when Thimmamma—a widow—committed *sati,* a Hindu practice where a woman immolates herself on her husband's funeral pyre. Her sacrifice gave life to the banyan tree, which sprung from the wood of the pyre.

It is also from these trees that the island of Barbados gets its name. When Portuguese explorer Pedro a Campos reached the island in 1536, he saw a great many banyans—in this case *F. citrifolia*—growing along the coast, the roots hanging from the trunks like great clumps of hair. He named the island *Los Barbados*—the bearded ones.

The banyan is so ubiquitous in India, and so beloved for the shade that it provides to villages and trade roads, that it has become just as central to the country's mythos as it is to daily life. According to Hindu mythology, the material world is described as a tree, the roots of which travel upwards and the branches of which travel down. This is Asvattha, the bodhi fig tree, and is in fact a real tree located in Bodh Gaya, India. It was beneath this tree that Siddhartha Gautama Buddha attained enlightenment in the 5th Century, and its leaves are said to be the resting place for the god Krishna, who first described the tree thus in the *Bhagavat Gita*: 'There is a banyan tree which has its roots upward and its branches down, and the Vedic hymns are its leaves. One who knows this tree is the 'knower of the Vedas.'

In the Philippines, the banyan (known locally as *balete*) is the home of a multitude of spirits called *Diwata*. The term literally means 'deity', and the Diwata would be invoked to provide blessings for crops, health, and good fortune. When the Spanish conquered the Philippines, they could not understand the concept of worshipping so many powerful deities at once, and so these benevolent gods were reduced to being called *engkanto*, 'enchanted'—a catch-all phrase for any and all humanoid spirits, including sirens, vampires, and ancestor spirits.

Alongside the Spanish there also came a whole host of new creatures called *maligno* ('evil spirits', from the Spanish for 'maligned'). These soon became associated with the banyan's empty trunks, which the locals knew never to point at or refer to directly for fear of drawing the attention (and subsequent ill will) of these creatures. To the people of the Visayas region of the Philippines, these foreign Spanish spirits became known as the *dili ingon nato,* meaning 'those not like us'. One of them was the *duende,* a dwarf that appears in Iberian and Spanish folklore, and originates from *dueño de casa,* 'possessor of a house' in reference to its nature as a mischievous domestic spirit.

The islanders continued to make distinctions between these new Spanish spirits and their own local ones. One of the local creatures said to inhabit the banyan was the *tikbalang*, a bony humanoid with the head and hooves of a horse and uncannily long legs, to the extent that when it crouches down, its knees reach above its head. An archetypal mischievous spirit, it would wait inside the banyan trunk, and lead passing travellers astray or return them always to the same path no matter how far they might walk.* Like the pranks acted out by the *metsänpeitto* in Finland, the tikbalang can be countered by turning one's clothes inside out, or by asking permission out loud to pass through. A tale popular with the Tagalogs says that the tikbalang isn't actually a troublemaking spirit at all; it is a guardian of the elemental world, and turns around anyone who might walk too close to the doorways into that realm.

Another, more ancient type of creature that belongs to the banyan trees is known as the *Taotaomona*, meaning 'people before history', and tales of these headless spirits are told by the Chamorro people of the Mariana Islands, just east of the Philippines. Like many other banyan spirits, they can be easily offended and inflict bad luck on a person or place. Their particular brand of mischief includes pinching, kidnapping, and imitating voices. They can sometimes become attached to a human, making them ill, and only a visit to a witchdoctor can remove them. This attachment of a spirit to a human is similar to the European concept of 'ghost sickness', where spirits of the dead can become attached to those who have been in contact with corpses.

STRYCHNINE TREE: Strychnos nux-vomica

They poured strychnine in his cup
and shook to see him drink it up:
They shook, they stared as white's their shirt:
them it was their poison hurt.

A. E. Housman, *A Shropshire Lad LXII*

Infamous for its poison of the same name, the strychnine tree is native to India and Malaysia. It is also one of the oldest species of tree to remain largely unchanged; in 1986 a flower from *S. electri* was discovered preserved in amber, almost identical in appearance to modern specimens, and is thought to date back to at least 15 million years ago.

* Isabelo de los Reyes; *El Folk-Lore Filipino*

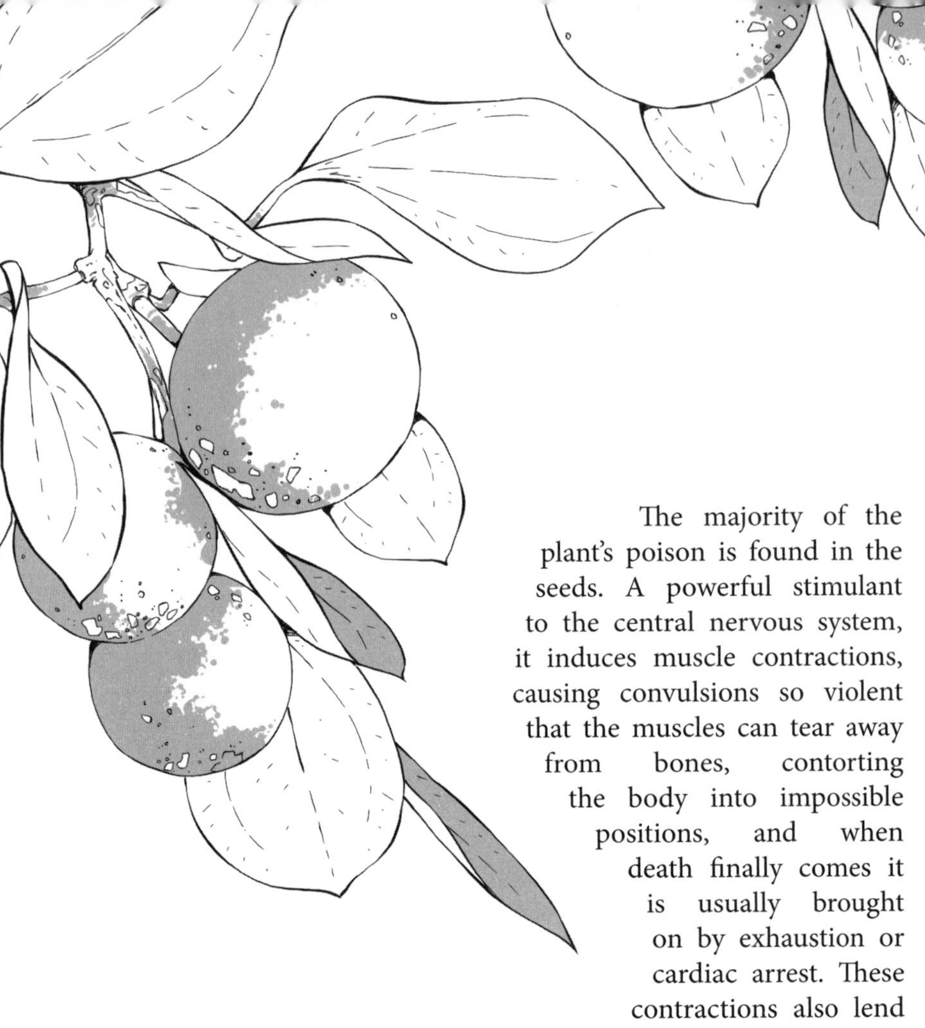

The majority of the plant's poison is found in the seeds. A powerful stimulant to the central nervous system, it induces muscle contractions, causing convulsions so violent that the muscles can tear away from bones, contorting the body into impossible positions, and when death finally comes it is usually brought on by exhaustion or cardiac arrest. These contractions also lend it the nickname 'the smiling poison', as it draws the lips back into a horrifying grimace.

Historically, it wasn't uncommon for people who feared death by poisoning to ingest small amounts of common toxins to build up an immunity; such was the case with Mithridates, who was famous for his work developing antidotes to poison. Unfortunately, no such method would work for strychnine: the body's sensitivity to it increases with repeat exposure, meaning that if small amounts were to be taken daily, what might begin as a relatively harmless dose could eventually prove deadly. For this reason, it was believed for a time that the body did not process strychnine but simply stored it away until a fatal level was reached, at which point the person would die of it.

The crystalline variety of the poison wasn't created until 1818, but the toxin itself from the seeds of the strychnine tree has been used for centuries. Two varieties of the plant, *S. tiente* and *S. toxifera,* yield poisons known in Java as *Upas Tiente* and in South America as *curare,* and are used to tip blowpipe darts, arrows, and spearheads. Strychnine's efficiency also saw the poison imported to Europe in the 15th Century as a pest killer, where it was used against rats, moles, and magpies. Although effective against mammals, it was remarked at the time that birds should be watched and culled quickly as the effects would simply make them drunk, and would rapidly wear off. It was still commonly purchasable as pest control as late as 1934 when it was responsible for the death of one Arthur Major, the unfortunate victim of his wife Ethel. His death was originally recorded as 'status epilepticus', or a prolonged convulsing seizure; but Ethel's downfall came when she left food scraps from the poisoned meal out for the neighbour's dog, who also died. Upon examining the animal, police discovered traces of strychnine and Ethel was convicted of murder.

Arthur isn't the only human to have fallen foul of this poison. Cleopatra, who famously ended her life to avoid humiliation at the hands of the military leader Octavian, first used slaves to test the effects of different poisons, experimenting with henbane, belladonna, and strychnine amongst others. Observing the distorting effects that strychnine had on the body, and wanting to ensure that she looked beautiful in death, she eventually chose a venomous snake to carry out the deed instead.

Strychnine was also the method chosen by infamous serial killer William Palmer, otherwise known as the Rugeley Poisoner. Though Palmer was only convicted in 1855 for the murder of his friend John Cook, it is suspected that he was also responsible for the deaths of his wife, five children, brother, mother-in-law, and another two friends. At the time of their deaths, they were thought to have died of cholera, stroke, alcohol, or cot death, but all were reassessed after Palmer's ultimate conviction for the death of Cook and subsequent attempts to bribe his way to freedom.

THORN APPLE: Datura stramonium

He gave her rich white flowers with crimson scent,
The tuberose and datura ever burning
Their incense to the dusky face of night.
He spoke to her pure words of lofty sense,
But tinged with poison for a tranced ear.
He bade low music sound of faint farewells,
Which fixed her eyes upon a leafy picture,
Wherein she wandered through an amber twilight
Toward a still grave in a sleepy nook.

George MacDonald, *Within and Without, Part V*

With toothed leaves and nocturnal white flowers that can grow up to eight inches across, the datura is a striking plant that became hugely popular as an ornamental specimen when it was introduced to the western world around 1550. Though it was originally thought that the plant could not have travelled outside of America prior to this, it must have been traded across the Pacific some time before as it also appears in Indian mythology, and even the common name comes from the Sanskrit *dhattura*, meaning 'poisoner'.* It is possible that this was enabled by the hardiness of the seeds—which can be viable for up to ten years—making it that much easier for the datura to have made its way across borders and over oceans.

* R Geeta and W Gharaibeh; *Historical evidence for a pre-Columbian presence of Datura in the Old World*

The thorn apple (so called for the spiky, round pods that protect the seeds) is related to the brugmansia, and the two are as remarkably similar in appearance as they are in toxicity. Yet another member of the Solanaceae family, datura is best known for its ability to cause sensations of drunkenness, laughter, and madness. Even just the scent of the flowers can cause hallucinations. In America it is known as Jimsonweed, as the early settlers of James Town in Virginia experienced these effects first-hand in 1679 when settling the land. A later record of this event talks of how the townsfolk, experimenting with the uses of the plants in this strange new land, ate the leaves and suffered multiple cases of madness and, ultimately, fatalities*. Seventy years later, during the first uprising against Britain, residents of the same town slipped it into the food of British soldiers, causing them eleven days of temporary insanity—although on this occasion, the soldiers were lucky and none of them died.

In 1802, when smoking was still a recommended form of relief for asthma sufferers, General William Gent brought back to England the leaves of *D. ferox* from Hindustan, where he declared that the locals used it to the same effect. However, he warned that smoking it could cause 'visions of convincing reality', some of which could linger for up to a week after use. This hallucinatory effect has led to a great deal of ill-fated experimentation by recreational drug users, with one account describing how they lost the ability to breathe automatically, and had to regulate their own breathing until the attack passed. This is likely because overdose can cause a failure of the autonomic nervous system, which regulates breathing and the heart. In cases where the plant has been used for divinatory purposes—such as by Native American tribes like the Californian Yokuts or the River Yumans of Colorado—centuries of use have resulted in a healthy understanding of and respect for its effects; dosages are strictly enforced and instances of overdoses are rare.

The Zuni in New Mexico use the datura to speak with the dead. They believe that the plant grows over portals to the realm of the ancestors, and came into the world through a pair of siblings who gifted a great deal of knowledge to humans. This brother and sister

* Robert Beverley; *History and Present State of Virginia*

came from the underworld, and one day followed the light upwards to the surface. Crowned with white flowers, they spent many years walking the earth, learning from the humans there and sharing what they knew. One day, they met the twin sons of the Sun Father, known as the Divine Ones. The brother and sister told them of their travels, and how they had taught humans how to see ghosts, how to sleep, and how to find lost objects. The Divine Ones decided that the children knew too much, and caused them to sink back into the underworld so that they would not be able to return. All that remained were the flowers they had worn in their hair, which grow even today, and continue the siblings' teachings.

The datura's toxicity plays an interesting part in the story of Haitian zombies. In places where the vodou religion is prevalent, zombies can be found working in bakeries, in fields and orchards, and guarding properties; they are people who have died and been brought back again as a slave. Only a sorcerer, called a *bokor*, can raise a zombie, and tales abound of entire plantations being worked by zombies, or of entire villages being massacred for

their bodies.* Zombies are an accepted part of the culture, and up to a thousand cases of zombification are reported annually.

It sounds fantastical, but zombies are entirely real, and it all boils down to a deep, cultural acceptance that zombification could happen to anyone. In 1985, Wade Davis, an ethnobotanist, visited Haiti to research exactly how *coupe poudre*—the 'zombie powder' that bokors use to create zombies—works.†

The main ingredient in the powder is a tetradoxin that comes from the puffer fish, an incredibly deadly poison. Though even a small dose can kill, an even tinier dose can reduce a person into a deathlike state, inducing bodily paralysis whilst keeping the mind alert; the self-same effect that puffer fish induce in themselves when threatened. Under the effects of this toxin, a person might be declared dead, and buried. When the bokor exhumes the 'body', they then administer datura and the plant *Mucuna pruriens,* the velvet bean, both of which cause hallucinations and amnesia. Under the effects of these substances, the person feels as though they have been revived in a dream-like stupor.

This is where the deep cultural belief in zombification comes into play. Many who have been born into cultures where vodou is practiced believe beyond question that escaping this fate, once it has befallen you, is futile; as a result, their mind accepts these alterations without question, causing them to bend more easily to the will of the bokor. Were the drugs to be given to someone from a different culture, the effect is not likely to be so complete—though the situation would, doubtless, remain a traumatising one.

However, against all odds, some zombies do manage to escape their situation, and through them we have a somewhat more complete understanding of the process. In 1962 a man named Clairvius Narcisse was admitted to a Deschapelles hospital with a fever, and after three days, died and was buried. Eighteen years later, he appeared on the doorstep of his sister, claiming to have been turned into a zombie, whereupon he was made to work on a plantation with other zombies.

* Francis Huxley; *The Invisibles: Vodoo Gods in Haiti*

† Wade Davis; *The Serpent and the Rainbow* and *Passage of Darkness: The Ethnobiology of the Haitian Zombie*

He remembered his own funeral, and had a scar on his cheek where a nail had been driven into his coffin. Around the time that he reappeared, several other people were found who recounted the same story, telling how they managed to escape when their master died. As the hallucinogenic effects of *coupe poudre* last less than a day, it would need to be reapplied regularly to maintain a state of zombification, and the death of the plantation owner would have likely broken the spell.

The datura has been noted in other parts of the world for its capacity for mind control. A 17th Century medical report of European origin claimed that the seeds, when swallowed by a man, will 'deprave and delude his mind to such a degree that anything can be done in his presence without fear of him remembering it on the following day. This madness of the mind lasts for twenty-four hours and you can do what you like with him; he notices nothing, understands nothing, and knows nothing about it on the next day.'* This effect is remarkably comparable to the use employed in Haiti.

A subspecies, *D. alba,* was used for similar purposes by the Thugs, an Indian sect of professional thieves and assassins. They held Shiva and Kali as sacred patrons, and dedicated the datura to Kali. They would serve a curry laced with the seeds (the curry disguising the bitter flavour) to travellers before robbing them. An account from 1883 of such a robbery is as follows:

Bassawur Singh, a professional Indian poisoner, ate some of the poisoned food to lull suspicion. In due course his victims fell insensible, and he robbed them, but after they came around and reported the theft to police, the thief was found about a mile away, quite insensible—and he never came around. All the stolen property was recovered, along with a supply of seeds.†

* Peter Haining; *The Warlock's Book: Secrets of Black Magic from the Ancient Grimoires*
† Alfred Taylor; *Principles and Practice of Medical Jurisprudence*

TSUBAKI: Camellia japonica

The cold Camellia only, stiff and white,
Rose without perfume, lily without grace,
When chilling winter shows his icy face,
Blooms for a world that vainly seeks delight.

Honore de Balzac, *The Camellia*

Camellia japonica, known best as the common or Japanese camellia (or *tsubaki* in Japanese) is one of the most beloved and best known of the camellia genus. These popular ornamental plants come originally from eastern and southern Asia, but are now cultivated and bred worldwide. The tsubaki is one of the original wild variants, and grows mostly in mountain forests. It flowers between January and March, usually during the snows, earning it the name 'the rose of winter'.

Held in much affection by the Japanese, the tsubaki has featured in both rituals and art for hundreds of years. A favourite of powerful figures such as Tokugawa Hidetada, the second Shogun of the Tokugawa Shogunate, it became a status symbol to grow or wear camellias, and in the Edo period (1603 to 1868) growing new domestic varieties became a popular pastime. One of the oldest shrines in Japan, the Tsubaki Okami Yashiro (the Tsubaki Grand Shrine) is named for the tree; built in the year 3BC, it is still in use to this today.

Even the dark, hard wood was considered a thing of beauty, and in 1961 archaeological excavations in Fukui prefecture discovered combs and axe handles made of tsubaki wood that date back approximately 5,000 years. The legendary emperor Keiko, thought to have ruled between 71AD and 130AD, was said to have killed his enemies with a hammer made of this wood, never once missing a blow.

The association with warriors continued with the rise of the samurai creed. The samurai are famed for being fond of cherry blossoms, their fleeting lives seemingly reflected in the short yet beautiful lifespan of the flowers; but so too were they often represented by the camellia, for similar—yet remarkably more brutal—reasons. The evergreen camellia doesn't lose its flowers gradually like the cherry,

but sheds them heavily and suddenly in one go, much like the fall of a severed head in the midst of battle. As a result, the plant has long been associated with death in Japan. This association is reflected in the *Hanakotoba,* the Japanese version of floriography—the Victorian language of flowers. The camellia represents 'perishing with grace', and a noble death.

A common belief in Japan states that when something (usually a non-sentient object) reaches an old age, it develops a spirit of its own and becomes a *yokai*. If it was treated badly during its lifetime, it can become vengeful and seek to punish those who mistreated it. The spirit of the camellia is called *furutsubaki no rei*, the old spirit of the tsubaki. One such yokai resides in the Kanman-ji temple in Akita prefecture, where the 700-year-old camellia that grows there is known as *yonaki tsubaki*, the night-crying tsubaki. The story behind it tells how one of the priests long ago heard a sad and lonely voice coming from the tree, and a few days later, a disaster befell the temple. Every time something bad would happen to the temple, the tree would have cried out the day before, warning them of the impending danger.

UPAS TREE: Antiaris toxicaria

But once a man dispatched a man
With one dread glance to that dead waste
And he obeyed. Away he ran
And brought the poison back with haste:
Its lethal sap, its waxen bough
And desiccated leaves. The sweat
Across his sallow, stricken brow
Ran in a chilling rivulet.
He brought it, stumbled and sprawled, prone
Beneath the tent for his reward:
A poor slave's death before the throne
Of his invulnerable lord.

Alexander Pushkin, *The Upas Tree*

Upas is the Javanese word for 'poison', so it's no surprise that a plant known as the upas tree should be toxic. There are two completely distinct plants that have been known historically under that name: the first is *Antiaris toxicaria*, a tree also known as anchar which can grow up to a hundred feet tall, and the second is the chetik,* a creeping shrub endemic to Java that creates a 'subtle and deleterious poison'.† Despite many references to chetik by botanists discussing the upas tree, as well

* Sir Thomas Stamford Raffles; *The History of Java*
† *History of the Indian Archipelago; The Edinburgh Magazine and Literary Miscellany, Volume 86*

as numerous descriptions regarding its biology, there has never been any scientific name recorded for the bush and it appears to no longer exist in the present day.

While the anchar and the chetik are indeed poisonous, they can't compare with the fictional upas tree, which became so famous that it engulfed the reputation of the real plant entirely. During the 17th and 18th Centuries, as naturalists and writers began to travel freely to newly-settled lands, tales of strange and wonderful new discoveries provided great potential for romantic dramatization and artistic license—and the upas tree was no exception. The most famous account was by John Nichols Foersch in 1773, whose stories gave it a reputation for being the most poisonous tree in the world.

Foersch claims that the tree emitted a gas so lethal that the land for fifteen miles around it was devoid of life, leaving the ground dry and barren. Even birds would refuse to fly over it. Yet the poison it created was so valuable that the emperor demanded it still be harvested—despite the fact that anyone approaching the tree had to do so with the wind behind them, and would need to wear leather gloves and a leather hood with glass eye holes to stand even a chance of survival. The men sent to do the task were generally criminals due for execution, incentivised by the promise that, should they survive the trip, they would be pardoned for their crimes. However, according to Foersch and the priest who showed him this spectacle, the survival rate was generally only one in ten.

A poem named *The Upas Tree* by famous Russian poet Alexander Pushkin tells a variant of this same tale. In this account, the upas tree grows in the 'wild and sterile' desert, and its poison, melted by the midday heat, leaks down its trunk in clammy drops. Like the original tale from Foersch, the air around the plant is saturated with the poison, and animals and birds won't go near it; but a slave is forced by his lord to gather the valuable toxin, despite the risk to his life.

It would be impossible for any tree to live up to such a fearsome reputation, though the poison of the real upas tree is perfectly deadly in its own right. There have been discussions as to whether there could be any truth behind Foersch's report; some have suggested that the tree he witnessed may have used allelopathy—secreting chemicals to kill off neighbouring competition—to clear the land around it. This behaviour is not uncommon in the plant world: the reed *Phragmites australis* uses acid to kill off the roots of other nearby plants, and the eucalyptus secretes an oil which, in hot weather, soaks into the ground to prevent the germination of rival seeds. However, Foersch's claim of a fifteen-mile radius is something of a stretch; though in 1837 W. H. Sykes suggested that the toxic fumes and the general barrenness of the land may have been the result of volcanic gas, which can kill via acidic erosion or asphyxiation, and could plausibly have covered such an area. Either way, it is almost certain that the toxic atmosphere was not the doing of the upas itself.

VIOLET: Viola spp.

Lilies for a bridal bed—
Roses for a matron's head—
Violets for a maiden dead.

Percy Shelley, *Remembrance*

Violets are small, edible wildflowers prized for their colour and scent. They grow mostly across the Northern Hemisphere, but have even been noted in Hawaii and the Andes, and a variety has been cultivated into the domesticated flower that we now call a pansy. The sweet scent of these flowers has likely contributed to their popularity, but a common myth warns that you will only ever be able to smell them once. Though not to be taken literally, there is an element of truth in the tale: ionine, one of the chemicals that contributes to the sweet scent, deadens the smell receptors for a period of time.

The genus name *Viola*, and the derivative for its common name, comes from the Greek nymph Ione (whose Latin name was Viola), the unfortunate object of Zeus's affections. Famous for his extramarital exploits, Zeus turned Ione into a white heifer to avoid the attention of his wife Hera. But Ione, despairing of this turn of events and the realisation that she must eat grass for the rest of her life, began to weep. Zeus took pity on her and turned the grass into violets, so that she might have something sweeter to sustain her.

Through both Greek and Roman literature, violets also became closely interlinked with mourning and death—a symbolism that still persists today. Violets would customarily be scattered around tombs, particularly those of children, and would be blanketed so thickly that the grave was often completely hidden by them. Even in the early 20th Century, when mourning dress was still common practice, violet was one of the colours worn in half-mourning after the period of full black had ended. This association may have come from Persephone, and the grief she bore at being taken by Hades into the Underworld; it was violets that she was gathering when she was taken, and that she missed the most when trapped beneath the ground.

Even Napoleon—who was personally fond of violets after his wife Josephine gave him a posy of the flowers upon their first meeting—used them as an expression of mourning for her death, being allowed to return just once to her grave before his exile to St Helena. Finding violets growing on her grave, he placed a few of these in a locket, which was found on him after his death. Even before this, his men used violets as a way to determine loyal supporters while he was in exile on the island of Elba. A stranger would be asked, 'Do you like violets?' If the answer was 'yes' or 'no', the person was clearly uninvolved in the plot to see Napoleon reinstated; but if they responded 'well…' then they were loyal.

The festival of Hilaria, meaning *joyful*, was a Greco-Roman celebration held in March in honour of the gods Cybele and Attis. The story of these gods is a dramatic one, and tells of the violent death and rebirth of Attis (who was renamed throughout other Mediterranean Basin faiths as Osiris, Tammuz, and Adonis). After a series of events that forced Cybele to drive Attis to madness, Attis castrated and killed himself at the base of a pine tree, before being resurrected by Zeus as a god. During the festival of Hilaria, which celebrated the ideals of death, mourning, and rebirth, a pine tree would be cut and placed in the temple, then wreathed in violets to symbolise Attis's tragic death. Novice priests of Cybele would then ritually recreate the scene at the foot of the pine, and emasculate themselves in sacrifice.

The association of violets with death can be found well outside of the Greco-Roman region, too. In Lithuanian folklore, the violet belongs to Poklius, the god of darkness and the underworld.* Similarly, the Prussian god Patulas, the god of death, was also often shown wreathed in violets; he is said to appear at night wearing chaplets of the flowers, with either the head of a dead man or a horse in place of his own.

* Jonas Lasickis; *Concerning the Gods of Samogitians, other Sarmatian and False Christian Gods*

WALNUT: Juglans regia

There upon antique marbles trac'd,
devices of past times we see,
here age hath almost quite defac'd,
what lovers carv'd on every tree.
The cellar, here, the highest room
receives when its old rafters fail,
soil'd with the venom and the foam
of the spider and the snail:
and th'ivy in the chimney we
find shaded by a walnut tree.

Katherine Philips, *La Solitude de St. Amant*

Well-established in both the Northern and Southern Hemispheres, walnut trees are well-known and prized for their brain-shaped nuts and smooth, dense timber. Though the majority of the world's walnuts are grown in California, they are used in dozens of industries across the globe. The husks make a dark yellow-brown dye popular in the fabric and wood trade, the plastics industry uses a flour made from walnut shells, and they are even used as a filler in dynamite!

Walnuts are commonly seen in parks and large gardens, as they are attractive trees that grow easily and quickly. But despite their popularity, they are not native to the English-speaking world. The

name gives it away: it was known in Old English as the *wealh-nut*, *wealh* meaning 'foreign'. Despite this, it has been growing outside of its native region of Iran since it was introduced to much of Europe by the early Romans.

The walnut isn't poisonous, but it is detrimental to other plants that grow near it. By way of allelopathy—the poisoning of one plant by another—it can stunt and kill certain plant species (others going unbothered, as they pose it no threat) that try to grow within 50 feet of its trunk. Allelopathy isn't unique to the walnut, and other plants that use it employ other tricks; some use acid to kill off competitors, while others rely on chemical-laden oils to inhibit root growth. The walnut uses a compound known as juglone, which deprives plants of the ability to metabolise, and eventually starves them to death.

For the longest time it was believed that humans could fall prey to the antisocial 'bad temper' of the walnut. The Romans believed that the shadow of the tree was particularly baneful, and could prove fatal; a later version of this same belief can still be encountered in Sussex, England, where it is thought that sitting or sleeping beneath a walnut can cause madness or even death. In Albania, when a walnut tree is too old to bear fruit anymore, it is said to have become haunted by a creature known as the *aerico*, a disease demon that originally came from Greece.

The Seven Sisters Road in Tottenham, London, is named for a grove of seven elms that grew in the nearby Page Green, first recorded in 1619. In the centre of this grove grew a single walnut, which was remarked in several publications to flourish every day yet never grow any larger.* The legend behind the sisters varies with each retelling, but one of the most popular versions says that the trees were planted by eight sisters, the youngest of whom planted an elm in the centre. When she was later burned under suspicion of witchcraft, the elm died too, and a walnut grew in its place. None of these trees exist anymore, the walnut having died in 1790 and the elms following in the mid-1800s, but they were more recently replaced with a ring of hornbeams in 1996, each one planted by a family that had been blessed with seven sisters.

The most infamous walnut on record is likely the Walnut of Benevento, which is said to be haunted by both the Devil and His witches. The story behind this revolves around Saint Barbatus, then a priest, who served the Church in the late 600s. An efficient hand at exorcism, Barbatus was sent to convert the people of Benevento, who worshipped a walnut upon whose trunk grew the appearance of a serpent. After convincing them to renounce their heathen ways, Barbatus uprooted the tree, and the Devil was seen in the form of a snake fleeing from beneath the roots. Though the courtyard where the tree grew is still empty, it is said that whenever the Devil calls for a sabbat, a walnut tree as large as the original appears on the same spot.

* Wilhelm Bedwell; *Brief History of Tottenham*

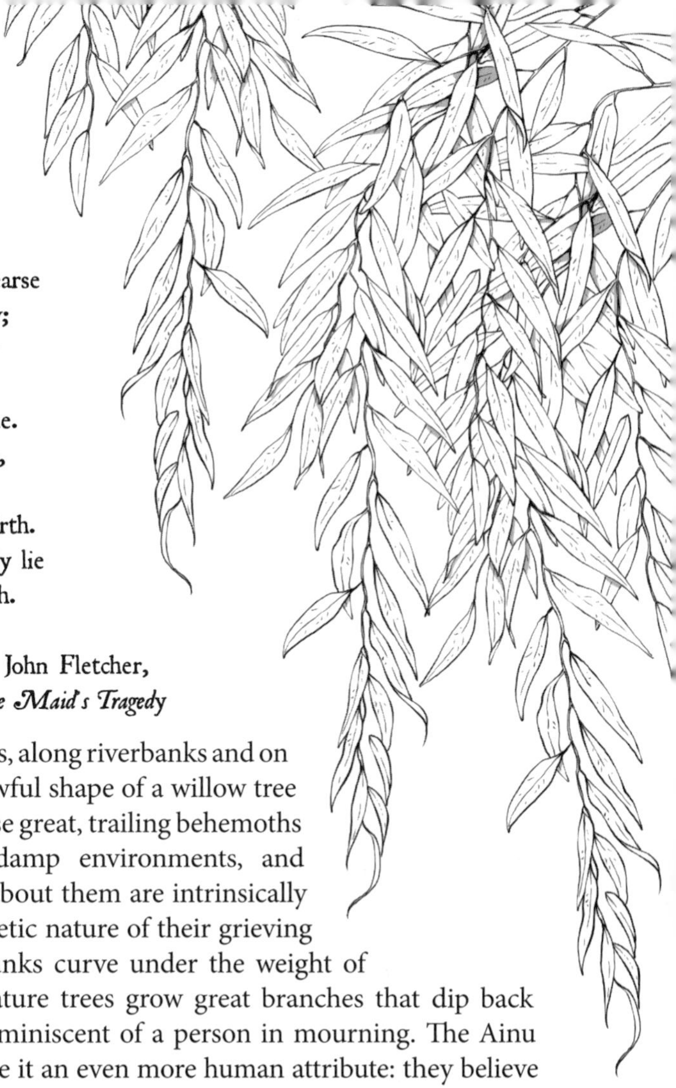

WILLOW:
Salix spp.

Lay a garland on my hearse
Of the dismal Yew;
Maidens, Willow
branches bear;
Say that I died true.
My love was false,
but I was firm
From my hour of birth.
Upon my buried body lie
Lightly gentle earth.

Francis Beaumont and John Fletcher,
Lay a Garland, from *The Maid's Tragedy*

At the edges of marshes, along riverbanks and on misty lakes, the sorrowful shape of a willow tree is a familiar sight. These great, trailing behemoths thrive in wet and damp environments, and many of the legends about them are intrinsically wrapped up in the poetic nature of their grieving forms. The young trunks curve under the weight of their boughs, and mature trees grow great branches that dip back down to the earth, reminiscent of a person in mourning. The Ainu people of Japan ascribe it an even more human attribute: they believe that the human backbone was originally formed of a willow branch. At the birth of every child, a willow tree is planted, and the child will continue to visit this personal guardian throughout his or her life, giving it beer and sake in exchange for their longevity.*

Beautiful and benign though they seem, the willow can—in some circumstances—be deadly. The bark contains salicylic acid, the main ingredient in painkillers, but the strength of the acid can vary greatly depending on amounts of sun, rain, and the quality of the soil

* John Batchelor; *The Ainu and their Folklore*

that the tree grows in. In too great a quantity, it can thin the blood to the extent that it may cause haemorrhaging.

Not only is the land in which the willow grows often eerie, particularly on misty mornings or dark nights, but the drooping branches and elongated leaves create a particular sort of whispering when disturbed by the wind. Rumours have long abounded that willows whisper to each other when they're alone; so it's advised not to speak any secrets near them! In the Czech Republic, a person who is untrustworthy with secrets is often called a 'hollow willow'.

The bending nature of willow fronds makes it easy to tie them into knots, and as a result, a form of knot magic has become unique to these trees. In Ireland, a person was able to ask a wish of a willow tree by speaking it whilst tying a loose knot in the branch. Once fulfilled, the same person would return to untie the knot. In Hesse, Germany, tying knots in the branches would place a killing curse on a victim;* but in England, tying a knot in a young willow was a way to renounce an unwanted baptism.† This kind of knot magic is usually done by the use of ropes, and dates back to early Egyptian and Greek sailors who would use it to bind the wind. Three knots would customarily be used: untying the first would release a gentle, south-westerly wind; the second, a strong north wind; and the third would release a tempest.

Various tales about the willow tree connect it to the same ghosts and supernatural creatures that haunt the wetlands it grows in. One such species of creature is the *russalki*, Slavic nymphs who live on sedimentary islands or in the coppices of trees on riverbanks. One story tells of a particular russalki who lived by day amongst humans, but always returned to her willow at night. She married a human, bearing his children and living happily with him; but one day he accidentally cut down her tree, whereupon she instantly died. Her son, however, continued to be able to communicate with her as he grew older by means of a pipe made from the wood of her tree.

* Thiselton Dyer; *The Folk-Lore of Plants*

† Cora Lin Daniels and Charles McClellan Stevens, *Encyclopaedia of Superstitions, Folklore, and the Occult Sciences of the World, Vol 2*

This legend isn't the only one that makes a connection between willows and music. An old Irish belief attests that the soul of the willow speaks through music, and many old Irish harps would be made from willow wood. The music of these instruments would supposedly inspire the uncontrollable urge to dance. According to the Christian Bible (Psalm 137), the boughs of the willow were originally straight, but became bowed when the Jews hung their harps upon them in Babylon, forever bending them downwards. Even Orpheus, during his ill-fated journey into the Underworld, carried willow branches with him—a nod to the muses, who were sacred to poets such as himself, and were referred to in *The Theogeny* as the Heliconian muses, after the willow nymph Helice.

Paramount in the lore that surrounds it, however, is the association between the willow and the concept of death. Particularly in Asia, where the tree thrives, it is considered a funereal tree greatly respected for its role in burial rites. In China, tombs and coffins are laid with branches of willow, which is seen as a symbol of purity, and the tree is often planted near the resting places of the departed. During Qingming, a festival in early spring when the dead are said to return to earth, willow switches are hung over doorways to ward off unwanted or uneasy spirits. In Japan, ghosts are said to be attracted to willow groves, and often appear near them.

In England, weeping willows were a popular image on Victorian mourning cards, or as gravesite ornamentations. However, a willow branch served a different purpose—a short, sharp stake was reserved for murderers and traitors, to be driven through the body after death to stop their angry spirit from returning to haunt the living. In Norfolk, until the 1800s, there grew a large willow tree that was reputed to have grown from one such stake. In Greece, too, the willow had a darker role: on Jason's voyage to find the Golden Fleece, he came across a grove of willows on the island of Colchis that was dedicated to Circe, the goddess of sorcery. These funereal willows wept lower than usual, thanks to the heavy weights of the corpses that hung from their branches.

WINGED CALABASH:
Crescentia alata

"Ah! What is the fruit of this tree? Is not the fruit borne by this tree delicious? I would not die. I would not be lost. Would it be heard if I were to pick one?" asked the maiden.
Then spoke the skull there in the midst of the tree:
"What is it that you desire of this? It is merely a skull, a thing placed in the branches of trees," said the head of Hunahpu.

The Popul Vuh, translated by Allen J. Christenson

The calabash tree is a small, nocturnal-flowering tree native to Central America. The blossoms are small, and grow directly from the trunk like flaps of skin; blooming only at night, they smell of carrion to attract nocturnal insects and bats. The fruits grow directly from the trunk as well, a feature known as cauliflory (literally 'stem flower'). The hard, cannonball-like gourds are difficult to break into, a defence mechanism against seed predation that is thought to have evolved long ago when megafauna inhabited the area. However, now that these large animals have become extinct the strategy has become counter-productive, as the fruit cannot germinate unless the shells are broken open—something that no native animals are equipped to do. However, domestic horses have been observed breaking the fruit open with their hooves, and may be responsible for the continued survival of the tree.

The gourds continue to have a use after harvesting, and are commonly repurposed as bowls, storage containers, and ornamental boxes. They are also used by the Taíno people of the Caribbean to hunt birds. Eyeholes are cut into the hollowed-out gourds, which are then worn on the head as the hunter enters a river or ocean. The birds are not afraid of the floating fruit, thus enabling the hunter to get close enough to drag them under the water without disturbing the rest of the flock.

The origins of the fruit of the calabash are written in the *Popul Vuh* (the 'Book of the People'), a work penned by anonymous members of the Quiché-Maya nobility in the 1500s which records centuries' worth of oral storytelling. The tale of the calabash tells of the first generation of hero twins, Hun-Hunahpú and Vucub-Hunahpú, and how the gods of Xibalba (the Mayan underworld) tricked them into losing a game of ball. The gods then beheaded Hun-Hunahpú and suspended his head from the branches of a calabash tree, and from the trunk, fruit that looked like Hun-Hunahpú began to spring forth where no fruit had ever been borne before. This tale is likely an explanation for the rotting smell of the tree and the skull-like shape of the gourds. Later in the story, a daughter of the Xibalban Lord Gathered Blood spoke with the head in the tree and became pregnant with the next generation of hero twins. These siblings went on to defeat the lords of Xibalba, and recover the remains of their father and uncle.

The Mayans aren't alone in having drawn a comparison between the human head and fruits on a tree. A Chinese name for the coconut tree (*Cocos nucifera*) is 'the head of the Prince Yue' after the legend of Yue, who was decapitated by an assassin whilst drunk. His head was hung on a palm tree and there it transformed into a coconut with eyes in the shell. In the Pacific Islands, where the coconut also grows, ritual sacrifices were a common occurrence but fell out of fashion with the introduction of Hinduism and the nonviolent practice of *ahimsa*. The skull-like similarity of the coconut to a human head made it a favoured replacement for living victims.

WISTERIA: Wisteria spp.

Around the door of the homestead,
The sweet Wisteria vines,
And on the old oak in the yard
The clinging ivy twines.
There stands the grim old court-house,
And the Jail with dingy cells,
And on the Church the old town-clock
The fleeting moment tells.
Next I came to the old town Graveyard
And entered with silent tread,
And dropped a tear o'er the grassy grave
Of the peacefully sleeping dead.

Josephine Delphine Henderson Heard, *Retrospect*

There are few sights more beautiful than a fully-grown wisteria in flower. These woody, climbing vines are a popular addition to empty walls and the fronts of houses, and can quickly grow to cover large areas. The largest recorded wisteria (and incidentally also the largest blossoming plant in the world) can be found in Sierra Madre, California; dated to 1894, it covers one acre of land, and is so enormous that it pulled down the original building that it grew against. Despite its beauty, the wisteria is in the same family as the laburnum, and is just as toxic. All parts of the plant contain wisterin, which can cause dizziness, speech problems, nausea, and collapse.

The delicate purple flowers of the wisteria are particularly prized in East Asia. A popular spring motif in Japanese kimono and kanzashi (hair ornaments), their blooms were once associated with nobility as purple was a colour forbidden in the clothing of commoners.

It was also a popular symbol used on *kamon*, family crests similar to European heraldic devices, as the Japanese name for wisteria—*fuji*—means 'immortal'.

In Korea, wild wisteria is commonly found growing on the hackberry tree (*Celtis jessoensis*), a relationship explained in the legend of *Deungnamu*. In Silla (a kingdom that existed between 57AD and 935AD), there were two sisters who both loved a Flower Knight of the Hwarang military corps. When the knights were called to battle, the sisters went to visit him on his final night before leaving and came face to face, thus learning of one another's love for the same man. Unable to give him up but unwilling to destroy their love for each other as sisters, they threw themselves into a pond and in death turned into a wisteria tree, their separate bodies entwined together. When the knight returned from battle and learned what had happened, he threw himself after them, turning into a hackberry, so that all three of them might always be together.

WOLFSBANE: Aconitum napellus

There hath been little set down concerning the virtue of the aconite, but much might be saide of the hurts that have come thereby.

John Gerard, *Great Herball*

Wolfsbane is infamous for its lethal nature. It is interchangeably known as monkshood, as the shape of the flowers somewhat resemble the cowls of English monks; however, wolfsbane is likely to be the older name for the plant as it can be found occurring much earlier in the Anglo-Saxon form *wulf-bana.** More poetic sources call it the Queen of Poison, such a favourite was it in acts of homicide. The scientific name comes from the hills of Aconitus, the place where the Romans claimed Hercules dragged Cerberus from the Underworld during their battle. Where the three-headed guard dog's saliva fell to the ground, wolfsbane sprung forth from the earth.

The name wolfsbane itself likely hails back to its use as a pesticide, where meat would be contaminated with the juice of the plant and then left out for livestock predators such as wolves and leopards. The Ainu of Japan would tip their weapons with this when hunting bears, and the Aleuts of Alaska would utilize it in whale hunting; it took just one man in a kayak armed with a poison-tipped lance to paralyse the creature, thereby causing it to drown.

Every part of this plant is deadly, and death occurs within two to six hours of ingestion. The initial signs are gastrointestinal, followed by a sensation of ants crawling beneath the skin, numbness in the mouth and face, and pronounced weakness. Death comes by way of paralysis of the heart and lungs, leading to asphyxiation. In 1856 near Dingwall, Scotland, the inhabitants of the monastery there discovered the dangers of this plant first-hand: one of the servants mistook the plant for horseradish, and mistakenly grated the roots into a sauce. Two of the priests died, but others recovered. The Greek physician Nicander of Colophon, in *Alexipharmaca,* described the sensations of poisoning thus:

* Philip Miller; *The Gardeners Dictionary: Containing the Best and Newest Methods of Cultivating and Improving the Kitchen, Fruit, Flower Garden, and Nursery*

When one takes aconite, the drinker's jaws and the roof of his mouth and his gums are constricted by the bitter draught as it wraps itself about the top of the chest, crushing the man with evil choking in the throes of heartburn. The top of the belly is gripped with pain ... and all the while, the moisture drips from his streaming eyes; and his belly sore shaken throws up wind, and much of it settles below about his mid-navel; and in his head is a grievous weight, and there ensues a rapid throbbing beneath his temples, and with his eyes he sees things double.

It was easily to be found growing wild in the Mediterranean Basin, and therefore freely accessible to the Greeks and Romans for nefarious purposes. Ovid called it 'mother-in-law's poison', aptly named given that almost eight hundred years later Walafrid Strabo suggested in *Hortulus* that horehound was an ideal antidote against wolfsbane poisoning 'if you should find yourself poisoned by your step-mother'. In Rome, aconite became so common a culprit for dinner party homicides that, in 117AD, Emperor Trajan made it a capital offence to grow it within the city walls.

Wherever it grows, wolfsbane is connected with death, rebirth, and transformation through sorcery. In the Greco-Roman mythology it was responsible for the fate of Arachne, who challenged Athena (or Minerva, in some versions of the myth) to a weaving competition. When it became clear that Arachne's weaving skills were greater than Athena's, the goddess threw aconite over her in a fit of rage, and turned her into a spider.

Another unsurprising connection is wolfsbane's contribution to tales of lycanthropy. Just as it was used to dispose of marauding wolves, it is said to be able to drive away werewolves, or otherwise halt their transformation.* Yet at the same time, encountering it during a full moon is said to induce lycanthropy in those who touch it.

* Ian Woodward, The Werewolf Delusion

YEW: Taxus baccata

Old Yew, which graspest at the stones
That name the under-lying dead,
Thy fibres net the dreamless head,
Thy roots are wrapt about the bones.
And gazing on thee, sullen tree,
Sick for thy stubborn hardihood,
I seem to fail from out my blood
And grow incorporate into thee.

Alfred Lord Tennyson, *In Memoriam*

If you've ever walked through a Christian cemetery, you more than likely will have seen yew trees growing there. Dark and bushy with bright red berries, they stand sentinel over burial grounds across Europe, and are one of the longest-lived tree species—no doubt one of the reasons they have become so synonymous with immortality and resurrection. Numerous specimens easily number at least 2,000 years old, and the Fortingall Yew in Perthshire is reputed to have grown in place for approximately 9,000 years. It's surprisingly hard to date the age of a yew accurately; as it grows the branches eventually curve downwards, and once they touch the ground they grow a new trunk that is still part of the original tree, but takes over once the first trunk rots away. In this manner, the tree is practically immortal. In the Irish Ogham calendar it is said to represent 'the new year that is born from the old, the new soul sprung from ancient roots in a seemingly fresh new body.'*

Though they are most prevalent in Christian churchyards, the custom of planting them in places of burial predates Christianity entirely. It was originally an Egyptian custom, which was then adopted by the Greeks, then the Romans, and finally brought to Britain. Even before the Romans introduced it to the British Isles, it was already a revered tree of mourning for the Irish Celts: they believed that the fine roots of the yew tree would grow through the eyes of the dead to stop them from seeing back into the living world and yearning for their previous life. A similar Breton belief was that the roots grew through the mouth, freeing the soul for its next rebirth. It's easy to see how superstitions like this came about; the delicate roots spread easily through and around old bones, as proven in 1990 when an ancient yew in Hampshire, England, was uprooted by a storm, exposing the skeletons of no less than thirty burial plots tangled in its roots.

In Wales, the Nevern Church—which dates back to the 6th Century—is famous for its yew trees, which have been consistently 'bleeding' for as long as they've been planted there (approximately 700 years according to church records). Though it is not uncommon for yew trees to 'bleed' sap when damaged, the scars usually heal quickly and it is unusual for one to have been weeping for so long. Some of the legends that have arisen to explain it include innocent monks hanged for crimes they didn't commit, the location of the unmarked grave of a high king, or a general empathy for the state of the world.

And so it follows that the yew tree has been romanticised and woven into ghost stories throughout the centuries. One particularly grisly tale is set in Halifax, Yorkshire, where a clergyman fell

* Colin Murray and Liz Murray; *The Celtic Tree Oracle: A System of Divination*

in love with a beautiful local girl who spurned his advances. Angry at being refused, he cut off her head and threw it into a yew tree, where it rotted. There is no real ending or moral to this story, but perhaps it is a reflection on the poisonous nature of the yew. Every part of the tree—apart from the flesh of the fruits—is toxic, and without any real evolutionary benefit. The tree does not suffer from animals grazing on the leaves or bark, and yet even the smallest amount can cause death to livestock and humans. Death for death's sake, it seems—a fitting notion for the tragic end of the Halifax maid.

It is only the seeds within the berries—technically arils, as they are modified cones that carry only one stone—that benefit from some element of toxicity. The arils themselves are benign, and actually quite sweet, but the seed inside of it has a coating that induces vomiting. In this manner the stone is eaten by mammals, carried some distance from the tree to a location where a sapling might have enough space and light to survive, and then ejected.

For those unlucky enough to swallow and retain more than just the fruit, death comes swiftly, and mercilessly. Physician Nicander of Colphon in *Counter Poisons* describes the symptoms of yew poisoning thus:

Be quick with aid, when yew tree juice with pains.
With anguish thrilling potion whelms the veins.
The tongue is under swol'n; the lips protrude
in heavy tumours, with dry froth bedew'd.
The gums are cleft; the heart quick terror shakes
smit with the bane; the labouring reason quakes.

Nicander's assessment of the toxicity of the yew isn't exaggerated, but little of the tree can be mistaken for edible and therefore the likelihood of accidental human poisoning it is low. However, the historical insistence on its dangers led to many dark rumours: tales persist that sleeping in the shadow of the tree can cause sickness or death,* for instance, and it was even believed that wine could be made poisonous

* John Gerard; *Great Herball*

by keeping it in yew barrels.* This last one is particularly unlikely; Irish wine barrels are typically made of yew (Robert Graves calls yew 'the coffin of the vine') to no known detriment of the drinkers.

The tree does enjoy some place in more romantic legends. Cemeteries are the practical resting places for our dead, but it's undeniable that there's something beautifully romantic about them too—the idea of memorialising people whose stories we may never know, the tombs of the too-soon departed, the mournful shapes of gravestones rising from the early-morning mist. Many of the stories involving the yew revolve around doomed lovers and the bond that remains between them even after death.

One particular Irish legend is that of the bard Phelim, who, after his daughter Dierdre was born, received a prophecy that many bloody wars would be fought over her beauty. Wishing to avoid such bloodshed, Phelim made a deal with King Conor of Ulster, who promised to hide her away until she was of age and then wed her himself. However, as Dierdre grew she had no desire to be wed to an older man. Sneaking away from her prison, she met and fell in love with Naoise, a handsome young nobleman. Together they fled to Scotland, where they lived happily for many years. But the King, angry to be denied his bride, lured them back to Ireland, sparking a war between their two families that resulted in the deaths of many, including Naoise himself. Overcome with grief, Dierdre killed herself, and from her grave grew a yew tree. The branches grew and spread until at last they reached the place where Naoise was buried. Here, too, another yew had sprung up, and in this manner the two lovers were once again united.

This tale shares some similarity to the medieval legend of Tristan and Isolde, who were parted in death but joined again by the ivy that grew between their graves; and also that of Barbara Allen, who was reconnected with her lover William by a briar bush.

* Mrs C. F. Leyel; *The Magic of Herbs*

INDEX

T

U

V

W

Y

BIBLIOGRAPHY/ FURTHER READING

Addy, Sidney Oldall; *Household Tales with Other Traditional Remains*, 1895

Agrippa, Heinrich Cornelius; *Three Books of Occult Philosophy*, 1533

Allen, David, and Hatfield, Gabrielle; *Medicinal Plants in Folk Tradition: An Ethnobotany of Britain and Ireland*, 2004

Jesus Azcorra Alejos; *Diez Leyendas Mayas*, 1998

Andía, Juan Javier Rivera; *Non-Humans in Amerindian South America: Ethnographies of Indigenous Cosmologies, Rituals and Songs*, 2018

Andrews, Jean; *Peppers: The Domesticated Capsicums*, 1995

Arnaudov, Mihail; *Snapshots of Bulgarian Folklore*, 1968

Awolalu, J Omosade; *Yoruba Beliefs and Sacrificial Rites*, 1979

Baker, Margaret; *Folklore and Customs of Rural England*, 1974

Baigent, Francis and Millard, James; *A History of the Ancient Town and Manor of Basingstoke*, 1889

Barber, Paul; *Vampires, Burials, and Death: Folklore and Reality*, 1988

Batchelor, John; *The Ainu and Their Folklore*, 1901

Beckwith, Martha; *Notes on Jamaican Ethnobotany*, 1927

Bedwell, Wilhelm; *Brief History of Tottenham*, 1631

Bergen, F. D.; *The Journal of American Folklore Vol. 2*, 1889

Bennett, Jennifer; *Lilies of the Hearth: The Historical Relationship Between Women and Plants*, 1991

Beverley, Robert; *History and Present State of Virginia*, 1705

Beza, Marcu; *Paganism in Romanian Folklore*, 1928

Boguet, Henri; *Discours Exécrable des Sorciers/An Examen of Witches*, 1602

Borza, Alexandru; *Ethnobotanical Dictionary*, 1965

Bottrell, William; *Stories and Folk-Lore of West Cornwall*, 1880

Boyer, Corinne; *Plants of the Devil*, 2017

Breitenberger, Barbara; *Aphrodite and Eros: The Development of Greek Erotic Mythology*, 2007

Brighetti, A; *From Belladonna to Atropine, Historical Medical Notes*, 1966
Briggs, Katharine; *An Encyclopedia of Fairies*, 1976
Brook, Richard; *New Cyclopaedia of Botany and Complete Book of Herbs*, 1854
Brown, Michael; *Death in the Garden*, 2018
Browne, Ray; *Popular Beliefs and Practices from Alabama*, 1958
Burton, Robert; *The Anatomy of Melancholy*, 1621
Carleton, William; *Traits and Stories of the Irish Peasantry*, 1834
Carrington, Dorothy; *The Dream-Hunters of Corsica*, 1995
Chambers, Robert; *Popular Rhymes of Scotland*, 1826
Christenson, A J; *Popol Vuh: Sacred Book of the Quiché Maya People*, 2007
Clark, H. F.; *The Mandrake Fiend*, 1962
Coles, William; *Adam in Eden*, 1657
Corner, George; *The Rise of Medicine at Salerno in the Twelfth Century*, 1933
Cousins, William Edward; *Madagascar of Today: A Sketch of the Island, with Chapters on its Past*, 1895
Crescenzi, Pietro de; *Ruralia Commoda*, 1304—1309
Daniels, Cora Linn, and McClellan Stevans, Charles; *Encyclopaedia of Superstitions, Folklore, and the Occult Sciences of the World*, 1903
Dauncey, Elizabeth, and Larsson, Sonny; *Plants That Kill: A Natural History of the World's Most Poisonous Plants*, 2018
Davis, Wade; *The Serpent and the Rainbow*, 1985
Davis, Wade; *Passage of Darkness: The Ethnobiology of the Haitian Zombie*, 1988
de los Reyes, Isabelo; *El Folk-Lore Filipino*, 1889
Debrunner, Hans Werner; *Witchcraft in Ghana: A Study on the Belief in Destructive Witches and its Effect on the Akan Tribes*, 1961
Duffy, Martin; *Effigy: Of Graven Image and Holy Idol*, 2016
Dwelley, Edward; *Dwelley's Illustrated Scottish-Gaelic Dictionary*, 1990
Dyer, Thiselton; *The Folk-Lore of Plants*, 1889
Eberhart, George; *Mysterious Creatures: A Guide to Cryptozoology*, 2002
Emboden, William; *Bizarre Plants: Magical, Monstrous, Mythical*, 1974
Evans-Wentz, Walter; *The Fairy-Faith in Celtic Countries*, 1911
Fernie, William Thomas; *Herbal Simples Approved for Modern Uses of Cure*, 1895
Folkard, Richard; *Plant Lore, Legends, and Lyrics: Embracing the Myths, Traditions, Superstitions, and Folk-Lore of the Plant Kingdom*, 1892

Frazer, James; *Jacob and the Mandrakes*, 1917
Friend, Hilderic; *Folk-Medicine: A Chapter in the History of Culture*, 1883
Gårdbäck, Johannes Björn; *Trolldom: Spells and Methods of the Norse Folk Magic Tradition*, 2015
Gary, Gemma; *The Black Toad: West Country Witchcraft and Magic*, 2016
Gibson, Marion; *Witchcraft and Society in England and America, 1550-1750*, 2003
Gifford, George; *A Dialogue Concerning Witches and Witchcrafts*, 1593
Gillam, Frederick; *Poisonous Plants in Great Britain*, 2008
Gillis, W. T.; *The systematics and ecology of poison-ivy and the poison-oaks*, 1960
Ginzburg, Carlo; *The Night Battles: Witchcraft and Agrarian Cults in the Sixteenth and Seventeenth Centuries*, 1983
Gooding, Loveless, and Proctor; *Flora of Barbados*, 1965
Graves, Robert; *The White Goddess*, 2011
Grieve, Maude; *A Modern Herbal*, 1931
Guazzo, Francesco Maria; *Compendium Maleficarum*, 1608
Hageneder, Fred; *The Meaning of Trees*, 2005
Haining, Peter; *The Warlock's Book: Secrets of Black Magic from the Ancient Grimoires*, 1971
Harkup, Kathryn; *A is for Arsenic: The Poisons of Agatha Christie*, 2015
Harvey, Steenie; *Twilight Places: Ireland's Enduring Fairy Lore*, 1998
Hatsis, Thomas; *The Witches' Ointment: The Secret History of Psychedelic Magic*, 2015
Heath, Jennifer; *The Echoing Green: The Garden in Myth and Memory*, 2000
Henderson, William; *Folklore of the Northern Counties of England and the Borders*, 1879
Hill, Thomas; *Source of Wisdom: Old English and Early Medieval Latin Studies*, 2007
Hooke, Della; *Trees in Anglo-Saxon England: Literature, Lore and Landscape*, 2010
Humphrey, Sheryl; *The Haunted Garden: Death and Transfiguration in the Folklore of Plants*, 2012
Hurston, Zora Neale; *Tell My Horse: Vodoo and Life in Haiti and Jamaica*, 1938
Huxley, Francis; *The Invisibles: Vodoo Gods in Haiti*, 1969

Johnson, Charles; *British Poisonous Plants*, 1856
Johnson, William Branch; *Folk tales of Normandy*, 1929
Josselyn, John; *New-England's Rarities Discovered in Birds, Beasts, Fishes, Serpents, and Plants of That Country*, 1672
Kaufman, David B.; *Poisons and Poisoning Among the Romans*, 1932
Kennedy, James; *Folklore and Reminiscences of Strathtey and Grandtully*, 1927
Kingsbury, John; *Poisonous Plants of the United States and Canada*, 1964
Knowlton, Timothy and Vail, Gabrielle; *Hybrid Cosmologies in Mesoamerica: A Reevaluation of the Yax Cheel Cab, a Maya World Tree*, 2010
Kuklin, Alexander; *How do Witches Fly? A Practical Approach to Nocturnal Flights*, 1999
Kvideland, Reimund and Sehmsdorf , Henning; *Scandinavian Folk Belief and Legend*, 1988
Lane, Edward William; *An Account of the Manners and Customs of the Modern Egyptians*, 1836
Jonas Lasickis; *Concerning the Gods of Samogitians, other Sarmatian and False Christian Gods*, 1615
Lawrence, Berta; *Somerset Legends*, 1973
Lea, Henry Charles; *Materials Toward a History of Witchcraft*, 1939
Leland, Charles; *Gypsy Sorcery and Fortune Telling*, 1891
Leyel, C. F.; *The Magic of Herbs*, 1926
Lockwood, T. E.; *The Ethnobotany of Brugmansia*, 1979
Lopez, Javier Ocampo; *Mitos, Leyendas y Relatos Colombianos*, 2006
Mabey, Richard; *Flora Britannica*, 1996
Mac Coitir, Niall; *Irish Trees: Myth, Legend and Folklore*, 2003
Máchal, Jan; *The Mythology of all Races. III, Celtic and Slavic Mythology*, 1918
MacGregor, Alasdair Alpin; *The Peat-Fire Flame: Folk-Tales & Traditions of the Highlands & Islands*, 1937
MacInnis, Peter; *A Brief History of Poisons*, 2004
Marren, Peter; Mushrooms: *The Natural and Human World of British Fungi*, 2018
McClintock, Elizabeth, and Fuller, Thomas; *Poisonous Plants of California*, 1986

Philip Miller; *The Gardeners Dictionary: Containing the Best and Newest Methods of Cultivating and Improving the Kitchen, Fruit, Flower Garden, and Nursery*, 1731
Millspaugh, Charles Frederick; *American Medicinal Plants*, 1887
Mooney, James; *History, Myths, and Sacred Formulas of the Cherokees*, 1981
Muller-Ebeling, Claudia, and Ratsch, Christian; *Witchcraft Medicine: Healing Arts, Shamanic Practices, and Forbidden Plants*, 2003
Multedo, Roccu; *Le 'Mazzerisme' et le Folklore Magique de la Corse*, 1975
Murray, Colin and Murray, Liz; *The Celtic Tree Oracle: A System of Divination*, 1988
Murray, Margaret; *The Witch-Cult in Western Europe*, 1921
Otto, Walter; *Dionysus: Myth and Cult*, 1965
Parkinson, John; *Theatrum Botanicum*, 1640
Paterson, Jacqueline Memory; *Tree Wisdom*, 1996
Phillips, Henry; *Flora Historica*, 1829
Pollington, Stephen; *Leechcraft: Early English Charms, Plant Lore, and Healing*, 2008
Poole, Charles Henry; *The Customs, Superstitions, and Legends of the County of Somerset*, 1877
Porter, Enid; *Cambridgeshire Customs and Folklore*, 1969
Porteus, Alexander; *The Forest in Folklore and Mythology*, 2001
Pratt, Christina; *An Encyclopedia of Shamanism*, 2006
Prior, R. C. A.; *On the Popular Names of British Plants*, 1870
Raffles, Sir Thomas Stamford; *The History of Java*, 1817
Randolph, Vance; *Ozark Magic and Folklore*, 1947
Rätsch, Christian, Müller-Ebeling, Claudia, and Storl, Wolf-Dieter; *Witchcraft Medicine: Healing Arts, Shamanic Practices, and Forbidden Plants*, 1998
Rätsch, Christian, and Müller-Ebeling, Claudia; *Pagan Christmas: The Plants, Spirits, and Rituals at the Origins of Yuletide*, 2006
Ricciuti, Edward; *The Devil's Garden: Facts and Folklore of Perilous Plants*, 1978
Robb, George; *The Ordeal Poisons of Madagascar and Africa*, 1957
Russell, Claire and Russell, William Moy Stratton; *The Social Biology of the Werewolf Trials*, 1989
Schulke, Daniel; *Veneficium (Second and Revised Edition)*, 2012

Schultes, Richard Evans; *The Plant Kingdom and Hallucinogens Part III,* 1970
Schultes, Richard Evans; *Plants of the Gods: Their Sacred, Healing, and Hallucinogenic Powers*, 1998
Sébillot, Paul; *Le Folk-Lore De France: La Faune Et La Flore*, 1906
Šeškauskaitė, Daiva; *The Plant in the Mythology*, 2017
Seymour, St John; *Irish Witchcraft and Demonology*, 1913
Shah, Idries; *The Secret Lore of Magic*, 1972
Sibley, J. T; *The Way of the Wise: Traditional Norwegian Folk and Magic Medicine*, 2015
Simoons, Frederick; *Plants of Life, Plants of Death,* 1998
Skinner, Charles; *Myths and Legends of Flowers, Trees, Fruits and Plants,* 1991
Spence, Lewis; *The Magic Arts in Celtic Britain*, 1949
Spencer, Mark; *Murder Most Florid, Inside the Mind of a Forensic Botanist,* 2019
Standley, Paul and Steyermark, Julian; *Flora of Guatemala*, 1946
Stevens-Arroyo, Antonio; *Cave of the Jagua: The Mythological World of the Tainos*, 1988
Stridtbeckh, Christian; *Concerning Witches, and those Evil Women who Traffic with the Prince of Darkness*, 1690
Taylor, Alfred; *Principles and Practice of Medical Jurisprudence*, 1865
Thiselton-Dyer, William; *The Flora of Middlesex*, 1869
Threlkeld, Caleb; *Synopsis Stirpium Hibernicarum*, 1729
Tompkins, Peter, and Bird, Christopher; *The Secret Life of Plants,* 1974
Tongue, Ruth; *Forgotten Folk-Tales of the English Counties*, 1970
Toynbee, Jocelyn; *Death and Burial in the Roman World*, 1971
Trevelyan, Marie; *Folk-Lore and Folk Stories of Wales*, 1909
Turner, Nancy and Bell, Marcus; *The Ethnobotany of the Coast Salish Indians of Vancouver Island*, 1971
Turner, William; *A New Herball: Parts II and III*, 1568
Tynan, Katharine and Maitland, Frances; *The Book of Flowers*, 1909
Various; *A Collection of Rare and Curious Tracts Relating to Witchcraft in the Counties of Kent, Essex, Suffolk, Norfolk, and Lincoln, Between the Years 1618 and 1664*, 1838
Vickery, Roy; *A Dictionary of Plant-Lore*, 1995

Vickery, Roy; *Vickery's Folk Flora: An A-Z of the Folklore and Uses of British and Irish Plants*, 2019

von Humboldt, Alexander; *Cosmos: A Sketch of a Physical Description of the Universe*, 1845

Wade, Davis; *The Serpent and the Rainbow and Passage of Darkness: The Ethnobiology of the Haitian Zombie*, 1985

Wasson, Valentina Pavlovna; *Mushrooms, Russia, and History*, 1957

Watts, Donald; *Dictionary of Plant Lore*, 2007

Webster, David; *A Collection of Rare and Curious Tracts Relating on Witchcraft and the Second Sight*, 1820

Weiner, Michael; *Earth Medicine – Earth Foods: Plant Remedies, Drugs and Natural Foods of the North American Indians*, 1971

Wellcome, Henry Solomon; *Anglo-Saxon Leechcraft: An Historical Sketch of Early English Medicine; Lecture Memoranda*, 1912

Wells, Diana; *Lives of the Trees: An Uncommon History*, 2010

Westwood, Jennifer and Kingshill, Sophia; *The Lore of Scotland: A Guide to Scottish Legends*, 2009

Wilde, Jane; *Ancient Legends, Mystic Charms, and Superstitions of Ireland*, 1902

Wood, J. Maxwell; *Witchcraft and Superstitious Record in the South-Western District of Scotland*, 1911

Woodward, Ian; *The Werewolf Delusion*, 1979

Woodyard, Chris; *The Victorian book of the dead*, 2014

Wright, Elbee; *Book of Legendary Spells: A Collection of Unusual Legends from Various Ages and Cultures*, 1974